T0179504

Cognitive Radio

Cognitive Radio

Basic Concepts, Mathematical Modeling and Applications

Rajeshree Raut

Government College of Engineering Nagpur, Maharashtra, India;
ISTE, New Delhi, India

Ranjit Sawant

Government Polytechnic, Hingoli, Maharashtra, India;
ISTE, New Delhi, India

Shriraghavan Madbushi

Department of Electronics & Telecommunication Engineering,
St. Vincent Pallotti College of Engineering and Technology,
Nagpur, Maharashtra, India

CRC Press
Taylor & Francis Group
Boca Raton London New York

CRC Press is an imprint of the
Taylor & Francis Group, an **Informa** business

First edition published 2020
by CRC Press
6000 Broken Sound Parkway NW, Suite 300, Boca Raton, FL 33487-2742

and by CRC Press
2 Park Square, Milton Park, Abingdon, Oxon, OX14 4RN

© 2020 Taylor & Francis Group, LLC
First edition published by CRC Press 2020

CRC Press is an imprint of Taylor & Francis Group, LLC

ISBN: 978-0-367-36858-6 (hbk)
ISBN: 978-0-429-35310-9 (ebk)

Typeset in Palatino
by Lumina Datamatics Limited

Contents

Preface

Cognitive radio has been envisaged as a driving technology that can solve the problem of the spectrum crunch. It has witnessed an exponential growth in last decade and is still continuing to grow. This book is primarily intended for the readers who want to dwell deeper into all the aspects of cognitive radio. This book has been written in a user-friendly style and lucid language and introduces the reader to basic to advanced-level topics. The text is presented in a manner that contributes to the understanding of the general concepts and those specific to the implementation of cognitive radio technology. The text also provides fundamental treatment to many basic principles, practical and theoretical concepts. Besides, this text can serve as a torch bearer for research in the field of cognitive radio technology. An appendix on installation of GNU radio (open source software) is also provided at the end for the users to implement the real scenario of cognitive radio.

It is assumed that the reader already has preliminary knowledge of wireless communication, digital communication, probability theory, etc. This book contains a total of 12 chapters covering various aspects of the cognitive radio technology.

MATLAB® is a registered trademark of The MathWorks, Inc. For product information, please contact:

The MathWorks, Inc.
3 Apple Hill Drive
Natick, MA 01760-2098 USA
Tel: 508 647 7000
Fax: 508-647-7001
E-mail: info@mathworks.com
Web: www.mathworks.com

Acknowledgment

The authors would like to extend their heartfelt gratitude to Dr A.A. Ghatol, Ex-Vice Chancellor of Babasaheb Ambedkar Technological University (BATU), Lonere and Dr K.D. Kulat, Professor, Department of Electronics & Communication Engineering, Visvesvaraya National Institute of Technology (VNIT), Nagpur, who have always been a source of inspiration. Thanks are due to Dr M.S.S. Rukmini, Professor and Dean Student Affairs, Vignan's Foundation for Science, Technology and Research (Deemed to be University), Vadlamudi, Guntur District, Andhra Pradesh, Rajesh Bhambare, Awani Gaidhane, Shital Gunjal, Priti Subramanium, Oveek Chatterjee, Rahul Awathankar, and Madhura Kulkarni for their contributions. A special word of appreciation goes to Dr Madhavi Sawant for her patient and meticulous checking of the entire manuscript. The authors also want to thank Mr. Amit Sharma and Mr. Kartik Seth of Amitec Electronics Pvt. Ltd. Delhi (www. amitec.co) for permission to include some of GNU radio experiments from their software-defined radio (SDR) laboratory manual. For more details, Mr. Kartik Seth can be contacted on his email: kartik.seth@amitec.co.

The authors would like to express their heartfelt gratitude toward Shri Vilas Tayade and Mrs Swathi Shriraghavan Madbushi for their constant encouragement and support throughout the preparation of the entire manuscript.

Authors

Dr Rajeshree Raut, Associate Professor, Electronics & Telecommunications, and Dean Students Affairs, Government College of Engineering, Nagpur, is a graduate (1998) and postgraduate (2002) from Government College of Engineering, Amravati, and has done her PhD in Error Control Coding in year 2011. She has 20 years of teaching experience and 1.5 years of industrial experience. Her achievements include Nagpur City Mayor Award as the Women Achiever, March 2019, **ISTE National award for Innovative Research in Engineering, 2017, ISTE National Award for Best Engineering Teacher, 2013, RTM Nagpur University Best Teacher Award, 2011**, IIFS New Delhi, Shiksha Ratna Award, 2011, more than 75 Research publications, 4 SCI, 20 Scopus journals, around 150 citations, H-index 7, i-10 index 5, 5 published patents, around 50 expert talks in engineering colleges. She has also delivered expert talks and chaired international conferences in countries like Mexico, Dubai, and Malaysia. She is the advisory and editorial board member for many international journals. She is currently guiding three research scholars. Six research scholars have been awarded PhD under her guidance. She is the only lady to represent ISTE Maharashtra & Goa Section, as SMC (Section Managing Committee). Currently, she is Executive Council Member ISTE, New Delhi. She is the Fellow of IEI (Institute of Engineers India). She has authored a book on "Error Control Coding; For Performance Improvement in Cognitive Radio". The project guided by Dr Raut, titled "Design of PSK System in Cognitive Environment", **was awarded "Best Engineering College Project Award"**, for the Maharashtra State, February 2016 by ISTE. The sponsored projects successfully executed by her include projects from Amitec Electronics, Inotek Antennas Pvt. Ltd., and Cresolv Tech. Pvt. Ltd. in association with IID (Indian Institute for Drones). She is a well-known personality for **her work in cognitive radio communications**.

Dr Ranjit Sawant received his bachelor's and master's degree, both in Electronics Engineering from Shivaji University in Kolhapur and Amravati University in 1990 and 1995, respectively. He obtained his PhD from Sant Gadge Baba Amravati University, Amravati. He also holds an advanced diploma in Computer System Software & Analysis from Board of Technical Education Mumbai and was awarded gold medal for standing first in the order of merit. He is serving as Principal of Government Polytechnic Hingoli since July 2016. Dr Sawant has also worked in government polytechnic at Kolhapur and Miraj as the head of Department of IT Engineering, Department of Electronics Engineering, and Department of Computer Engineering.

He has also served in Government College of Engineering, Amravati, as lecturer for six and a half years. He also has industry experience of one and a half years as a trainee engineer in Machinen Polygraph [Manu Graph], at MIDC Kolhapur.

Dr Sawant is a fellow of IEI (Institute of Engineers India) and has worked as National Executive Council Member of ISTE, New Delhi, as secretary of Maharashtra and Goa section of ISTE, New Delhi. He has also served on different committees of Board of Technical Education (MSBTE), DTE (Director of Technical Education), Mumbai, and MSCIT (Maharashtra State Certificate in IT).

He has also delivered invited guest lectures in different engineering colleges in India and abroad and published various research papers at national and international level conferences and journals. He played a pivotal role in starting Diploma in Information Technology and BTech Electronics course of YCMOU.

He has also received a grant of rupees 15 lacs for setting up various IT labs from Maharashtra Vikas Vaidhanik Mandal and has to his credit the "Achiever" award in workshop conducted by Advanced Institute of Management, Dubai.

Dr Shriraghavan Madbushi holds bachelor's degree in Electronics Engineering from RTM Nagpur University (erstwhile Nagpur University), Nagpur, Maharashtra, and master's in Electronics & Communication Engineering from University College of Engineering, Osmania University, Hyderabad, Telangana. He obtained his PhD in Electronics & Communication Engineering from Vignan's Foundation for Science, Technology & Research (VFSTR), Vadlamudi, Guntur District, Andhra Pradesh. Dr M. Shriraghavan is currently working as Assistant Professor in the Department of Electronics & Telecommunication Engineering at St. Vincent Pallotti College of Engineering and Technology, Nagpur, Maharashtra, India. His areas of interest include cognitive radio and machine learning for wireless communications.

1

Introduction

1.1 Introduction

Cognitive radio (CR) and opportunistic spectrum sharing are promising ideas for boosting the proficiency of spectrum usage, which is otherwise vacant for most of the time. Both of them require an alternate level of discernment about the encompassing condition and an alternate level of refinement, which prompts distinctive difficulties. CR is a hopeful technology that is anticipated as an impending candidate that can assuage the problem of spectrum crux. This is promising in a scenario where unlicensed (secondary) users concur with the incumbent (primary) user in licensed spectrum bands without inducing any intrusion to the incumbent communication. One of the essential mechanisms to achieve this is *spectrum (range) sensing*. To alleviate the problem of scarcity of spectrum the Federal Communications Commission (FCC) has allowed the use of spectrum bands by the unlicensed users as far as they do not cause any interference to the primary users. Also, the secondary users have to vacate the spectrum as soon as they detect any incumbent user communication. This opportunistic use of spectrum increases the efficiency of the spectrum utilization and is referred to as *opportunistic spectrum access* (OSS) [2].

A cognitive radio (CR) is a radio that can change its transmission parameters in light of the apparent accessibility of the range groups in its working condition. CRs support dynamic spectrum access and encourage an auxiliary unlicensed client to productively use the accessible underutilized spectrum designated to the essential authorized clients.

A cognitive radio system (CRN) is made out of both the optional clients with CR-empowered radios and the incumbent users. A CR is characterized as a radio that can change its transmitter parameters according to the requirements and is cognizant about the environment in which it works. A CR has the capacity (intellectual ability) to detect and assemble data (e.g. the transmission recurrence, transfer speed, control, balance, etc.) from the encompassing condition and has the capacity (reconfigurability) to quickly adjust the operational parameters, for ideal execution, as per the data detected. With the above highlights, the CR innovation is being seen as the key empowering innovation for the cutting edge dynamic spectrum access systems that can proficiently use the accessible underutilized spectrum allotted by the Federal Communications Commission (FCC) to authorized holders, known as essential clients. CRs encourage a more adaptable and extensive utilization of the constrained and underutilized range for the optional clients, who have no range licenses.

CRs empower the utilization of transiently unused range, alluded to as range gap or blank area, and if an essential client means to utilize this band, at that point the optional client ought to consistently move to another range opening or remain in a similar band, adjusting its transmission control level or balance plan to abstain from meddling with the essential client. Customary range assignment plans and range get to conventions may never again be relevant when optional unlicensed clients exist together with essential authorized clients. In the event that auxiliary clients are permitted to transmit information alongside essential clients, the transmissions ought not to meddle with each other past an edge. If optional clients can transmit just without essential clients, at that point, auxiliary client transmitting information without an essential client ought to have the capacity to distinguish the return of the essential client and clear the band. At present, there is a lot of research being conducted and more should be performed to grow new range of administration approaches identified with CR for both range detection and dynamic range sharing.

Intellectual radio system engineering incorporates segments relating to both the auxiliary clients (optional system) and the essential clients (essential system). The auxiliary system is made out of an arrangement of optional clients with or without an optional base station, which are all furnished with CR capacities. An auxiliary system with a base station is alluded to as the framework-based CR. The base station goes about as a center point gathering the perceptions and consequences of spectrum investigation performed by every optional client and settling on the most proficient method to maintain a strategic distance from intrusion with the essential systems. According to this choice, every auxiliary client reconfigures the correspondence parameters. An auxiliary system without a base station is alluded to as the foundation less subjective radio specially appointed system or cognitive radio ad-hoc network (CRAHN). In a CRAHN, the CR auxiliary clients utilize collaboration plans to trade privately watched data among the gadgets to expand their insight into the whole system and choose their activities in light of this apparent worldwide information. An essential system involves essential clients and at least one essential base station, which are all as a rule not furnished with CR capacities. Consequently, if an auxiliary system shares an authorized range band with an essential system, the optional system is required to be capable to recognize the nearness of an essential client and direct the optional transmission to another accessible band that would not meddle with the essential transmission.

1.2 Need for CR

In numerous groups, spectrum access is a more major issue than physical shortage of spectrum. Because of heritage summon and control direction, the capacity of potential range clients to get such access is restricted. By checking the parts of the radio range, we find that:

1. Some recurrence groups in the range are very empty more often than not.
2. Some different groups of recurrence are incompletely possessed.
3. The rest of the groups of recurrence are to a great extent utilized. Spectrum hole is that in which a specific group of frequencies are relegated to a specific client at some particular land locale for some particular time, and this band is not utilized by another client around them. The CR empowers the utilization of unused range, which is known as void area or range opening. When the primary user is not utilizing the particular band, then the secondary user can utilize that band and enhance the spectrum usage resulting in the effective utilization of spectrum. The unused spectrum band is referred to as the spectrum hole. Figure 1.1 shows the illustration of spectrum hole.

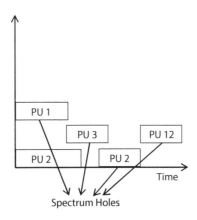

FIGURE 1.1
Spectrum hole.

1.3 Characteristics of CR

The main features of CR are:

1. It can decide its geographic area.
2. It can recognize and approve its clients.
3. It can perform encryption and decryption.
4. It can detect adjacent radios.
5. It can adapt to the environment in which it operates.

Figure 1.2 shows the cognition cycle on which the cognitive radio works [1]
Figure 1.3 shows a simplified version of the cognitive cycle.

Observe: A CR is a gadget that has four wide inputs; to be specific, a comprehension of the environment in which it works, a comprehension of the correspondence prerequisites of the client, a comprehension of the system, and the administrative arrangements that are applied to it, and its very own comprehension abilities. In the end, a CR knows about the setting in which it is working.

Decide: A CR gathers data about the spectrum usage from different sources and reconfigures itself and decides the best way for the correspondence jobs that needs to be done. In choosing how to arrange itself, the radio is cognizant about the imperatives or clashes (physical, administrative, etc.) that may exist.

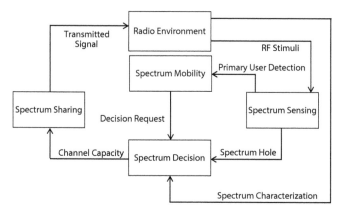

FIGURE 1.2
The cognition cycle.

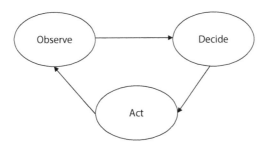

FIGURE 1.3
Simplified cognitive cycle.

> *Act*: A CR is produced using programming and hardware parts that can encourage the wide range of designs it needs to impart.

A CR is based on what is called a software defined radio (SDR) [3]. The software allows the radio to tune to different frequencies, power levels, and modulation depending upon its learning and the environment in which it operates. The CR is expected to perform the following four functions:

1. *Spectrum Sensing*: detection of "white spaces" or the portion of spectrum vacant for use. This must also ensure that no primary user (licensed user) is operating at the same time.
2. *Spectrum Management*: selection of the best spectrum hole for transmission.
3. *Spectrum Sharing*: sharing of spectrum with other potential users.
4. *Spectrum Mobility*: vacate the band when a licensed user is detected (spectrum handoff).

The functions mentioned above form a cognition cycle that forms the basis on which the CR operates. The cognitive cycle consists of five states, namely, Observe, Orient, Plan, Decide, and Act.

1.3.1 Cognitive Ability

It is the capacity of the radio innovation to detect the data from its radio condition. There are some modern procedures that are required to catch the fleeting and spatial variety in the radio condition rather than basically be acknowledged by observing the power and maintain a strategic distance from different clients. By this ability, the unused range can be distinguished at a particular time and area. However, the best range and suitable working parameters can be chosen. The subjective capacity empowers ongoing collaboration with its climate to decide appropriate correspondence parameters and change in accordance with the dynamic radio environment. The three stages of cognitive cycle under this classification are:

Spectrum detection: An intellectual radio recognizes the accessible range groups, takes their data, and then identifies the range gaps.

Spectrum examination: The spectrum gaps that are distinguished by the spectrum detection are evaluated.

Spectrum choice: A subjective (cognitive) radio breaks down the information rate, the data-transfer capacity of the transmission mode. At that point, the appropriate spectrum band is selected as indicated by the range qualities and client necessities. Once the spectrum band is identified, the correspondence can be started by this band.

1.3.2 Reconfigurability

Range mindfulness is chosen by the cognitive ability; then again, reconfigurability advises the radio on how to program the radio condition powerfully. The radio can transmit and tune to various frequencies to utilize distinguishing transmission parameters. Reconfigurability is the ability of altering working parameters for the transmission on the fly, with no changes on the equipment segments. There are several other reconfigurable parameters, which are as follows:

Operating frequency: Operating frequency can be changed by radio. The most reasonable working recurrence can be resolved and the correspondence can be progressively performed on the proper working recurrence based on the data about the radio condition.

Broadcast power: Among the power requirements, transmission power can be reconfigured. Power control empowers the dynamic transmission control arrangement inside the passable power confine. There is no need of high-power task; at bring down level, the intellectual radio enables more clients to share the range and diminish the intrusion.

1.4 Types of CR

Depending on the set of parameters taken into account in deciding on transmission and reception changes, and for historical reasons, we can distinguish certain types of CR.

Full CR ("Mitola radio"): In this radio, every possible parameter observable by a wireless node or network is taken into account.

Spectrum Sensing CR: In this radio, only the radio frequency spectrum is considered. Also, depending on the parts of the spectrum available for CR, we can distinguish:

Licensed Band CR: in which CR is capable of using bands assigned to licensed users, apart from unlicensed bands, such as U-NII band or ISM band. The IEEE 802.22 working group is developing a standard for wireless regional area network (WRAN) that will operate in unused television channels.

Unlicensed Band CR: this can only utilize unlicensed parts of radio frequency spectrum. One such system is described in the IEEE 802.15 task group 2 specification, which focuses on the coexistence of IEEE 802.11 and Bluetooth.

1.5 Underlay, Overlay, and Interweave Transmission

CR brings an innovative idea that permits respected coordinated innovation that permits opening up of the spectrum holes to simultaneous working clients in a nonmeddling mode. Likewise, to make conceivable spectrum sharing without making hazardous intrusion to the existing traffics, cognitive clients ought to have at least some data about their encompassing noncognitive clients. Contingent upon the information that is expected to exist together with the primary system, CR broadcast fall into three classes, Overlay, Underlay, and Interweave.

1.5.1 Overlay Transmission

In this approach, the cognitive devices may improve and assist the noncognitive broadcast rather than competing for spectrum. This is possible because the noncognitive users share information of their messages and signal codebooks with the cognitive users. More precisely, cognitive devices eavesdrop to the messages sent by noncognitive sources and make use of these messages either to eradicate the intrusion engendered by the primary communication at the cognitive receiver side or to improve the performance of the primary transmission through conveying the amassed messages to the primary receiver. The latter case allows the cognitive device to communicate at the same time as the noncognitive transmitter on the condition that its overall transmit power is fairly covering the energy requirements of its own communication as well as its transmitting operation. A trade-off should be judiciously designed amid the intrusion induced on the primary signal and the improvement brought to it to attain a stagnant SNR.

1.5.2 Underlay Transmission

Concurrent cognitive and noncognitive broadcasts are allowed as long as the obstruction level at the incumbent user side remains tolerable. Beyond the predefined endurable intrusion threshold, the primary signal may degrade dramatically. In recent literature, many advanced signal processing techniques have proven to be very efficient for interference avoidance and mitigation, including beam forming and spread spectrum. Beam forming consists of exploiting the superposition concept of waves to guide the signal toward a specific receiver using multiple antennas. More importantly, in a cognitive context constructive or destructive interference is provoked at the intended cognitive receiver in order to lessen the interference caused to non-cognitive users while focusing on the signal energy in the direction of secondary users. Using the spread spectrum technique, the secondary signal is multiplied by a spreading code to obtain a weaker signal with a wider band. The resulting spread signal causes a lower interference level to noncognitive users. The original secondary signal is recovered at the receiver side by simply multiplying the input signal with the same spreading code. The spread spectrum technique is also useful for alleviating the interference caused by the primary signals to the secondary ones. Another common solution could be limiting the power of the secondary signal to keep the interference level at the primary side bounded albeit restricting the secondary transmissions to short-range communications.

1.5.3 Interweave Transmission

This spectral coexistence has been projected with the intent of enabling devices to take up the spectrum rooms that have been left unoccupied by noncognitive users. The neighboring milieu should be observed to be able to envisage the condition of every segment of the frequency spectrum; portions of spectrum that are well thought out as being underutilized may be accessed by secondary users as long as the primary activity remains inactive. In order to make possible the coexistence of both primary and secondary traffics in the same network in an opportunistic broadcast mode, spectrum opportunities should be dynamically recognized and monitored.

On the one hand, cognitive users may carry out sensing operations permanently and, dependably, diverse dimensions have to be investigated to find the copious spectrum hole. Legacy sensing algorithms scrutinize and oversee the spectrum through three conventional dimensions: frequency, time, and space domain. Nevertheless, other degrees of freedom (DoF) such as the used code and the angle of arrival may be inspected. On the other hand, geographic coordination through a central database to recognize the unoccupied gaps is a fine substitute for self-centered spectral sensing. Merging both methods may be also envisaged.

Hybrid schemes using an amalgamation of the aforesaid paradigms have an immense potential to perk up the competence of spectrum allotment. The advantage of such schemes is that they permit secondary users to exploit their broadcast rate once a spectrum prospect is perceived.

1.6 Advantages of CR

CR is a new idea in the area of wireless sensor network that can make use of the spectrum frequency competently. CR systems have the potential like packet loss diminution, power waste diminution, buffer supervision, and enhanced communication. Following are the advantages of CR in wireless sensor networks:

1. *Resourceful utilization of spectrum*: The spectrum bands that are accessible can be used proficiently as these cannot be augmented. For exploiting the radio bands in a country, a permit is obligatory from the administration perspective. This is a pricey course of action. There are assured licensed spectrum bands that are under-utilized. CR can exploit this spectrum, and permit holders are also not bothered. This unutilized spectrum is known as *white space*.

2. *More space for new-fangled technologies*: New-fangled technologies can be developed for these vacant bands. Cost would also be negligible when this vacant spectrum is used by an unlicensed user.

3. *Ability to make use of manifold channels*: In a customary wireless sensor network, a solitary channel is used for communication. In this, when a happening is detected, the sensor nodes produce packets. In a crowded network, sensor nodes attempt to engage the solitary communication channel at the similar time. This augments the likelihood of crash, and the communication superiority is diminished. Furthermore, there is additional power expenditure and packet holdup. CR gives the chance to make use of manifold channels, thus dipping the chances of collision and at the same time augmenting the communication quality.

4. *Energy competence*: In customary wireless sensor network, the power expenditure is additional owing to augmented packet retransmission owing to packet loss. This energy expenditure due to packet retransmission is trounced by CR.

5. *Worldwide operability*: There are convinced spectrum directive rules for each country. Each country has a prearranged dissimilar spectrum band that is not presented in another country. CR gives the competence to be operated in any country of the world.

6. *Exploitation of spectrum band that is application precise*: There is an augment in wireless sensors that are set up for diverse applications. The wireless sensor network generates packets when an event is triggered. The data traffic in wireless sensor networks is associated temporally and spatially, which causes design confront of the communiqué etiquette. There are intelligent communiqué protocols in CR wireless sensor network that can conquer the design challenge of protocols.

7. *Financial benefit*: Those users who cannot obtain a permit for spectrum due to financial difficulty can get hold of this on lease from permit holder at a low cost. This will be beneficial to both users.

8. *Assaults are shun*: CR uses a wide range of spectrum; therefore, several types of attacks can exist, unlike in other wireless sensor network that works on a particular band of frequency.

1.7 Applications of CR

Military and security: CR is used in a variety of military and public services like chemical, biological, and nuclear radiation detection, war surveillance, and so on. Conventional wireless sensor networks face the predicament of signal jam. However, this predicament no longer exists if we use CR wireless sensor network. CR can handoff a wide range of frequencies. It is also helpful when the applications call for a hefty bandwidth.

Healthcare: CR also unearths its applications in the healthcare. It finds its use in wearable body wireless sensors. By means of this, significant information from patients can be obtained by doctors sitting at a far-away location. These wireless sensor nodes are very steadfast.

Home appliances: Wireless sensor network is used in certain indoor applications. However, there are little challenges in those indoor applications. CR trounces these challenges. The examples of these applications are intellectual buildings, home supervising system, and private amusement.

Real-time applications: In multi-hop wireless sensor setup, there are odds of link breakdown that can cause holdup in communication. Furthermore, the nodes move to an additional channel if they come across a different inactive channel. CR wireless sensor setup increases the channel bandwidth, making channel aggregation and use of manifold channels probable.

Transportation: CR comes across its applications in a transport system as well as vehicular system. A lot of exploration work is going on in this field, with a variety of code of behavior proposed for highway security. These are also credible to be additionally dependable and expedient than the customary wireless sensor network.

Assorted sensing: CR wireless sensor setup can be used in the identical area but with a diverse object. CR can use dissimilar channels for diverse applications using medium access protocol (MAC).

1.8 Challenges in CR

CR wireless sensor setup is different from customary wireless sensor setup in certain aspects. Even though they are way enhanced than these conventional wireless sensor networks, they also face some challenges. Some of these are as follows:

Misdetection probability and false alarm: These are the two metrics used in CR wireless sensor setup. It is a metric used in CR wireless sensor when it fails to become aware of the primary signal's attendance on the channel and false alarm is a metric for CR wireless sensor when it fails to sense the nonattendance of the incumbent signal. These two metrics can cause tribulations like long waiting delay, recurrent channel switching, and ruining output.

Hardware issues: There are certain constrictions applied on hardware of the system like computational power, energy, and storage. To devise a smart hardware for CR is an arduous assignment. The hardware should be such that it should be competent to react to the changes in the milieu. The basic philosophy of CR can be satisfied using certain artificial intelligence techniques like artificial neural network, Markov models, case-based systems, and so on. More or less, it can be said that the design of the hardware is an awfully demanding issue.

Changes in topology: The topologies in CR are more prone to change than the conventional wireless sensor network. This can cause hardware malfunction and diminution of energy. Following are the topologies employed in CR wireless network:

Ad Hoc

Clustered

Heterogeneous

Hierarchical

Mobile

Quality of service: Quality of service is a demanding issue due to certain restriction on power, memory, and source. The QoS is determined by four parameters—bandwidth, delay, jitter, and reliability. Consequently, satisfactory level of QoS is desirable to be preserved.

Fault tolerance: Fault tolerance is the most demanding problem in CR-WSN. The protocols intended for CR do not have the competence to endure fault. This has an effect on the overall performance of the wireless sensor setup.

Selection of channel: In CR-WSN, the channel is not fixed for sending data and for communication. This can cause non-cooperation amid the nodes. This challenge can be overcome by designing some AI algorithms.

Power consumption: More energy is consumed for spectrum sensing, channel assortment, route detection, broadcast, and reception than for customary setup. It also needs to sense the node's commotion on the channel.

Security: CR-WSNs are also more prone to a variety of solitude and defense issues than the customary WSNs. An assailant can attack the sensors and capture the not-to-be-disclosed information. These data can also be ruined by hackers.

High manufacturing costs: High manufacturing cost is an additional confront for CR wireless setup. The conventional WSN entail less memory and computation than the CR-WSN. As a result, algorithms need to be designed that can lower the manufacturing cost and production.

1.9 Simulation Tools of CR

Major concepts of CR have been handled all the way through dissimilar writing studies and the development of trustworthy expository models. Nonetheless, numerical re-enactments must be performed to verify the hypothetical outcomes for reasonable and exact discoveries.

Practical reproductions have been performed utilizing numerous PC-supported re-enactment devices. Specifically, the software products such as Matlab, Simulink, and C/C++ have been of extraordinary help to the research fraternity.

To handle the CR idea from a physical layer perspective, GNU radio is an intense, free, and open source software dedicated to programming the

SDRs. It tends to be matched with a Universal Software Radio Peripheral, which is an equipment intended to work for remote radio frequency frameworks. It can be additionally modified utilizing LabVIEW toolbox to model propelled framework setups including single-channel, MIMO, and embedded frameworks.

NetSim is likewise a standout among other instruments that have been utilized to survey the execution of the CR standard IEEE 802.22. It has a superior and simple-to-utilize user interface and simple source C records that can be altered for various troubleshooting purposes. From documentation point of view, network and online bolster are clear and simple to discover.

SEAMCAT assesses different potential impedance programming instrument. It assesses different potential impedance utilize cases between various radio correspondence advances utilizing Monte Carlo based knowledge. SEAMCAT permits considering for the most part the range detecting highlight among the equivalent or the neighboring recurrence channels; in such case, the meddling clients attempt to recognize the nearness of secured administrations as indicated by a prefixed location limit. It has been utilized to research the impedance between auxiliary frameworks working in TVWS and office-holder recipients to be specific, DVB-T and PMSE for the COGEU venture in Europe, and gave astute ends.

Recently, CorteXlab is a trial lab facilitated at INSA-Lyon in France created with the goal of giving remote access to an exploration stage with full test conceivable outcomes and substantial scale subjective radio test beds.

1.10 Conclusions

This chapter discusses CR technology and the vital role it plays in making the better use of spectrum to fulfill the large demand for wireless applications, ranging from intelligent grid, safety for public, and broadband cellular, for health applications. The transmission modes of underlay, overlay, and interweave have also been discussed in brief. Applications and challenges of CR have been discussed in brief. At the end, an overview of the simulation tools has been presented in brief.

Several regulatory establishments have begun to build up standards to take benefit of the opportunities. Nevertheless, challenges still remain since CR-enabled frameworks have to coexist with primary as well as secondary users and need to pay no attention to intrusion in such a way that they can better support such applications from end to end.

References

1. J. Mitola, III, and G. Q. Maguire. 1999. Cognitive radio: Making software radios more personal. *IEEE Personal Communication*, 6(4): 13–18.
2. J. Mitola, III. 2000. *Cognitive Radio: An Integrated Agent Architecture for Software Defined Radio*. PhD Diss., Royal Institute of Technology, Stockholm, Sweden.
3. F.K. Jondral. 2005. Software-defined radio—Basics and evolution to cognitive radio. *EURASIP Journal on Wireless Communications and Networking*, 3: 275–283.

2

Software-Defined Radio

2.1 Introduction

A software-defined radio (SDR) is a radio in which some or the entire physical layer functions are software defined [1]. Various definitions can be found to depict SDR, otherwise called software radio. Essentially, SDR is characterized as a radio in which a few or whole physical layer capacities are programming characterized. In SDR radio functionalities such as signal processing, signal generation, modulation/demodulation, and signal coding is implemented in software rather than the hardware as in conventional radio systems. The software realization results in reconfigurability and high amount of flexibility. Further, this results in other advantages such as modification in broadcast parameters, change in communication protocol, ability to alter channel assignments, and communication services. In short, SDR is considered as the technology enabler for cognitive radio (CR). SDR is also called as a software radio. The SDR dates back to the time when programming was first utilized inside radio and radio innovation.

The essential perception of the SDR is that the radio can be completely defined or configured by the software so that a universal platform can be used across a number of areas. The software is used to alter the design of the radio as per the need at a specified time. It is also possible that the radio can be upgraded and reconfigured as new standards develop or the scope of operation is changed. One noteworthy activity that uses the SDR, programming-characterized radio, is a military endeavor known as the Joint Tactical Radio System (JTRS) [2]. This could be utilized as a solitary equipment stage, and it could impart utilization of one of the assortment of waveforms just by reloading or reconfiguring the product for the specific application required. This is an especially appealing recommendation, particularly for alliance style tasks where power from various nations may work together. Radios could be rearranged to enable communication between the troops from various nations.

The SDR idea is similarly appropriate for the business world too. One application might be for cell base stations where standard redesigns happen much of the time. For instance, the transfer from UMTS to HSPA and on to LTE could be obliged just by transferring new programming and reconfiguring it with no equipment changes, in spite of the fact that diverse modulation schemes and frequencies might be utilized. In spite of the fact that it might sound an unimportant exercise, making a definition for the SDR is not as straightforward as it appears. It is additionally important to deliver an appropriate definition for some reasons including administrative applications, benchmarks issues, and empowering the SDR innovation to move advances more rapidly.

Numerous definitions have created the impression that may cover a definition for an SDR. The SDR Forum has characterized the two fundamental sorts of radio containing software in the accompanying design:

Software-controlled radio: Radio in which some or all of the physical layer functions are software controlled. In other words, this type of radio uses only software to provide control of the various functions that are fixed within the radio.

SDR: Radio in which some or all of the physical layer functions are software defined. In other words, the software is used to determine the specification of the radio and what it does. If the software within the radio is changed, its performance and function may change. Another definition that appears to envelop the quintessence of SDR is that it has a nonexclusive hardware equipment stage on which software runs to give capacities including filtering, modulation, and demodulation and other different activities such as frequency hopping, frequency selection, etc. By reconfiguring, the execution of the radio is changed at that point.

To accomplish this, the SDR innovation utilizes programming modules that keep running on a nonexclusive hardware equipment comprising of advanced digital signal processors (DSPs) and in addition general-purpose processors (GPPs) to actualize the radio capacities to broadcast and receive signals. The principle points of interest of SDR over conventional radio interchanges are:

- A solitary SDR gadget can play out different capacities basically by changing programming modules.
- System updates can be actualized in programming to be downloaded by means of the transmission system. These incorporate updates to both the product application and to any delicate configurable equipment.
- Highly configurable flag preparing frameworks can be created with alterations and upgrades made far less difficult to execute than more customary DSP frameworks.

- More adaptable correspondences conventions can be created that adjust to their condition straightforwardly to the framework client (e.g., hunting down and working in locally accessible groups). The idea of CR showed up as another worldview in 1999 as an expansion of SDR. It depicts the circumstance where insightful radio gadgets and related system substances impart in such a way that they can alter their working parameters as per the necessities of the client/system. From that point forward, there has been a lot of exertion in the examination network on CR-related subjects. Institutionalization exercises on Cognitive Radio Systems (CRSs) (counting TV WhiteSpaces—TVWS) have additionally been started, in addition to advances in numerous institutionalization bodies. All administrative bodies in the USA, Europe, and Asia-Pacific locales have recognized the significance of CRS on forming the manner in which range is distributed. Controllers like FCC in the USA and Ofcom in the UK have opened the entryway for auxiliary access to unlicensed gadgets on TV groups.

2.2 Wireless Innovation Forum Tiers of SDR

It is not always viable to build a radio that integrates all the features of a completely SDR. A few radios may just help various highlights related with SDRs, although others might be completely programming characterized. With the end goal to give a wide categorization of the radios, the SDR Forum (now called the Wireless Innovation Forum, WINNF) has defined a number of tiers. These tiers can be elucidated in terms of which part of radio is configurable.

- *Tier 0*: A nonconfigurable equipment radio, that is, it cannot be changed by programming.
- *Tier 1*: Limited functions such as interconnections, power levels are controllable. However, mode or frequency cannot be controlled.
- *Tier 2*: A significant portion of radio is configured via software. Therefore, most of the times the term *SCR* is used. The parameters controlled by software include waveform detection/generation, modulation, frequency, security, wide/narrow band operation, etc. However, the RF front end still is realized in hardware and is nonreconfigurable.
- *Tier 3*: This tier resembles Ideal Software Radio or ISR. Here, the edge amid configurable and nonconfigurable elements exists extremely close to the antenna, and the "front end" is configurable. ISR is said to have full programmability.
- *Tier 4*: This tier is one step ahead of tier 3 that is the ISR, and is known as the Ultimate Software Radio (USR). USR has full programmability and is also able to sustain a wide range of functions and frequencies at the same time.

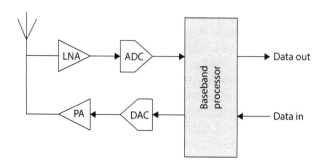

FIGURE 2.1
Block diagram of an ideal SDR (www.cablefree.net).

Despite the fact that these SDR levels are not binding at all, they give a method for extensively condensing the distinctive levels of programming characterized radios that may exist (Figure 2.1).

2.3 SDR Architecture

The hardware for an SDR is a predominantly significant element of the overall design. Although the radio is essentially driven by software, it still requires indispensable hardware to facilitate the software to run. The hardware development engineer faces some fascinating challenges while devising the SDR hardware. The performance of the hardware will characterize exactly how much can be done inside the software. The equipment for an SDR is an especially imperative component of the general outline. While the entire thought of the radio is that it is essentially determined by programming, regardless it needs the fundamental equipment to empower the product to run. An SDR is in general an assortment of hardware and software technologies where a few or all of the radio's working functions are put into operation through flexible software or firmware operating on programmable devices such as the GPPs, DSPs, system on chip, field programmable gate array (FPGA), or any other explicit programmable processors. These technologies facilitate innovative features and wireless capabilities to be integrated into radio systems devoid of any need to alter any hardware [3].

The indispensable building blocks of an SDR are shown in Figure 2.2. It consists of a real-time operating system, a software framework that makes available basic libraries and functions to support waveforms and their portability collectively with the middleware. Software framework is an amalgamation of Common Object Request Broker Architecture (CORBA) middleware and software communications architecture (SCA). The waveform represents the

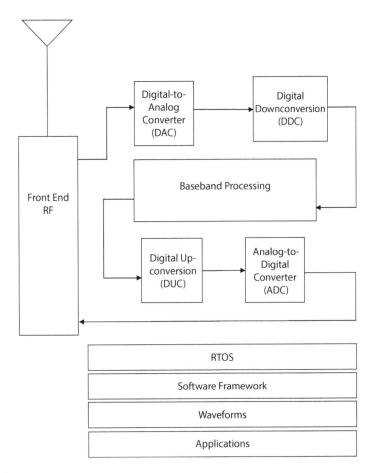

FIGURE 2.2
Simple SDR architecture. (From Baldini, G. et al., *IEEE Commun. Surv. Tut.*, 14, 355–379, 2012.)

software implementation of communication service such as UMTS. At the end, applications can be defined to support a particular business or operational framework. The transmitter and receiver structure of SDR is discussed in more details below. Along with the above parts. digital-to-analog converter (DAC), analog-to-digital converter (ADC), digital upconversion (DUC), and digital downconversion (DDC) blocks are also used.

2.3.1 The Transmitter Structure of SDR

Figure 2.3 depicts diagram of SDR transmitter. It comprises of DSP, DUC, DAC, analog RF upconverter, and power amplifier. The DSP is used to code the digital baseband part that provides data as per diverse transmitter needs. A DUC is then used to digitally upconvert the same with the help of a digital mixer

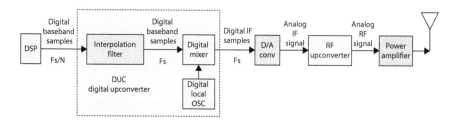

FIGURE 2.3
Block diagram of SDR transmitter (www.rfwireless-world.com).

and a digital logic oscillator. The digital IF samples are transformed to analog IF. A RF upconverter then converts the analog IF to analog RF. The RF signal is amplified prior to be broadcast over the air by means of an antenna [3].

2.3.2 The Receiver Structure of SDR

The first block is RF tuner as shown in Figure 2.4. The RF tuner translates the RF signal to amplified IF signal. ADC converts analog IF into digital IF samples [3].

The digital samples are then passed to the DDC, which then converts digital IF samples into digital baseband samples and are referred as I/Q data. DDC comprises of a digital local oscillator (LO), digital mixer, and a low pass FIR filter.

The digital baseband samples are then accepted by the DSP chip. Here, the algorithms have been ported. The diverse functions performed at this stage include demodulation, decoding, and any other such processes if required.

This digital realization based design is referred as SDR. Frequently, FPGA is also used as a substitute to DSP in software-defined design. The exquisiteness to have baseband processing chain realized in software on DSP/FPGA will help in acceptable real-time baseband and RF linked impairments present in I/Q data with the utilization of classy algorithms. Algorithms such as I/Q gain and phase imbalance correction, time, frequency, and channel impairment correction, DC offset correction are put into practice in SDR receiver.

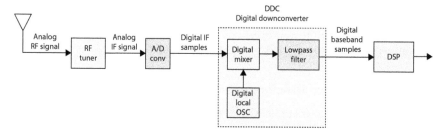

FIGURE 2.4
Block diagram of SDR receiver (www.rfwireless-world.com).

2.4 Benefits of Using SDR

The benefits of SDR are convincing. They are discussed one-by-one as follows:

1. *Reconfigurability*: In SDR, the entire functionality and interface can be reconfigured to suit the user requirements. This is done by changing the software alone without the need to change the hardware.

2. *Reusability*: Once the software has been tested and developed, it can be reused on other hardware platforms. This can also be used to create new products or entire product families, thereby saving the time and cost of developing new products.

3. *Reduced cost*: Owing to the software reusability and decline in the number hardware components, it is easier and cheaper to fabricate SDR products. A significant portion of the costly RF hardware is substituted by general-purpose and economical computer hardware. As a result, the cost of SDR is also falling, whereas its performance keeps on improving to set latest standards.

4. *Stability and consistency*: Hardware components are susceptible to changes in temperature, manufacturing variations, aging, wear and tear, and so on. This will lead to degraded performance of the components. However, this is not the case with software.

5. *Easily upgraded*: Upgrading the hardware is not an easy task. It involves changing the components, boards, modules, etc. In contrast, software can be easily upgraded by simply changing a piece of software or few lines of the code adding new functions and features. This further enhances the use of the product extending its lifetime.

6. *Flexibility in design*: The tremendous progress in the SDR technology is due to the availability of available computer power, sophisticated signal processing algorithms, and easy availability of accurate and fast ADCs. The software has moved closer to the antenna as more and more customary hardware components are replaced by software. It is possible to perform filtering, demodulation, and other functions in software.

2.5 Portability of SDR Waveform

Aside from the way that the SCR can reconfigure itself, another significant preferred standpoint is waveform portability. There are a few explanations behind the requirement for SDR waveform portability. The waveforms for various military and commercial applications can be used on different projects and this is likely to involve very different platforms, thus saving cost. Whenever there is a need to change hardware as the technology develops, it is obligatory to reassign on hand waveforms on newer platforms. To give finished interoperability, a client may ask for the utilization of a specific waveform being utilized over the equipment gear from a few manufacturers. SDR waveform portability is not at all times easy to accomplish. However, it is indispensable to integrate measures at the most primordial stages of the design to guarantee the optimum level of portability. Techniques that may work on one platform are undoubtedly not likely to work on a new one. It is frequently required to compile again the code for use in diverse platforms; hence, all codes should be in a format that can be compiled again and again.

2.6 Testing of Interoperability of SDR

With the need to transfer waveforms from one radio or platform to another, it is compulsory to embark on full interoperability testing. This needs a guarantee that the code can be effortlessly transported from one platform to another. This helps to provide the acceptable functionality for the particular waveform, and this would be achieved when the waveforms are by and large endorsed and authorized.

SDR is a reality today, and it is being utilized in numerous applications. Anyway, there are various confinements that are anticipated. One is the absolute processing power that is required, and the subsequent power utilization. It is important to embrace a power utilization/handling power transaction, and this is one of the center choices that should be made at the beginning. Likewise, programming-characterized radios are being utilized by the military, and right now some handheld designs are also foreseen. As innovation advances, programming-characterized SDRs will be utilized in more applications; yet, a choice needs to be made, as the SDR is not the correct choice for all radios. For little modest radios in which changes will be few, the SDR is unsuitable. However, for more confounded frameworks in which the length of administration is an issue and where change is likely, the SDR is undoubtedly a decent alternative to be considered

2.7 SDR Security

SDR security is another zone of developing significance. Numerous military radios and frequently numerous business radio frameworks should guarantee that the transmissions stay secure, and this is an issue that is critical for a wide range of radio. Anyway, when utilizing a product-characterized radio, SDR, there is another component of security, in particular, that guarantees that the product inside the radio is safely downloaded. With the development of utilization of the Internet, numerous SDRs will utilize this medium to convey their updates. This shows an open door for malevolent programming to be conveyed that could alter the task of the radio or maintain its activity by and large. In like manner, SDR programming security should be considered if the Internet is utilized for programming conveyance or where there could be security shortcomings that could be utilized malignantly.

2.8 The Software Communications Architecture

The SCA is a product standard that is utilized related to the SDR. The SCA gives a typical standard that intends to give compactness of a few components, and a typical interface with the goal that diverse modules composed by various gatherings can be united effectively [4].

The thought of the SCA is the brain child of the JTRS. Amid the venture, it was important to gather programming for the SDR from various distinctive providers. It was additionally important to have the capacity to reutilize programming wherever conceivable. As needs be, the SCA was characterized and executed.

SCA is utilized for overseeing the structure and the task of the product inside an SDR, empowering them to stack waveforms, run applications, and be arranged into a coordinated framework. The utilization of SCA additionally gives enhanced levels of interoperability among radio sets.

2.9 Basics of SCA

SCA fundamentally portrays the product segments inside an SDR and specifically characterizes the interfaces [4]. The utilization of SCA gives two fundamental points of interest:

1. It empowers programming components or modules to be composed by various associations and to be united.
2. It empowers the reutilization of a few modules, consequently enhancing interoperability and giving huge cost reserve funds.

SCA programming can be categorized into three categories. It is valuable to classify them, since they should be dealt with some unique ways, and some can be reutilized over a few stages while others may not. The three SCA classes are discussed below:

1. *Management*: Software that falls into this SCA class is used for supervision of the radio framework. An assortment of applications may comprise of plug and play; deployment and configuration software.
2. *Node*: This software may consist of applications such as bootstrapping and right to use the hardware.
3. *Application*: This sort of software is used predominantly for signal processing, which takes into account demodulation, frequency translation, waveform generation, and so on.

Of these three diverse SCA classes, application programming can be reutilized between various stages. The other two need to exclusively interface with the hardware and are platform precise.

2.10 Common Object Request Broker Architecture

Middleware known as CORBA is used to the make possible intermodule communications. Its use is a vital part of the SCA standard. It is a key component in empowering modules and components of programming written by various associations to be united. CORBA is standard software defined by the Object Management Group. The intent is to facilitate software modules written in an assortment of different computer languages and running on diverse platforms to be brought collectively and function [4].

2.11 SCA Testing and Acquiescence

Altogether, any product that can be pronounced as SCA programming should be tried for SCA acquiescence. Along these lines, the Applications Programming Interface can be resolved as agreeable, and it will work with other SCA accommodating software. Additionally, its execution is tried for right task. SCA is a perfect standard to be used for expansive SDR ventures where distinctive programming components might be acquired from various programming houses. It gives a hearty interface between the distinctive programming modules that enables parts to impart together dependably in a known standard organization. However, the utilization of SCA places an overhead on the intricacy of the framework. This may imply that SCA may not be the correct decision for smaller tasks. Regardless of whether to utilize SCA or not is an outline decision that should be made at the start of the venture [4].

2.12 Conclusions

This chapter gives a fundamental overview of SDR development. Further, discussion on transmitter and receiver structure is presented. SDR portability, testing, and operability issues are also discussed in brief. A concise discussion on security is also presented. At the end, deliberation on vital SCA is presented in brief.

References

1. J. Mitola III, 1992. Software radios—Survey, critical evaluation and future directions, in *Telesystem Conferencing* 1992. NTC-92, National, May 1992: 13/15–13/23, IEEE, Washington, DC.
2. F. K. Jondral, 2005. Software-defined radio—basics and evolution to cognitive radio, *EURASIP Journal on Wireless Communications and Networking* 2005(3): 275–283.
3. https://www.rfwireless-world.com/Articles/SDR-Software-Defined-Radio-basics.html.
4. https://www.wirelessinnovation.org/.

3

Source Coding for Audio and Speech Signals

3.1 Introduction

Speech and audio coding is widely used in applications such as digital broad casting, Internet audio, or music database to reduce the bit rate of high-quality audio signal without comprising the perceptual value. Techniques have also been emerging in recent years that offer enhanced quality bit rate over traditional methods. Wideband audio compression is generally aimed at a quality that is nearly indistinguishable from consumer compact-disk audio. Sub-band and transform coding methods with sophisticated perceptual coding techniques dominate in this area with good quality bit rates.

Compression of audio signal has found application in many areas, such as multimedia signal coding, high fidelity audio for radio broadcasting, and audio transmission for HDTV, audio data transmission/sharing through Internet, and so on. Speech compression converts input speech data stream into a reduced size data stream by eliminating inherent redundancy associated with speech signals. Compression techniques reduce the overall program execution time and also storage requirement of processor. Compression helps to reduce the data transfer rate and bandwidth requirement with security for data to be transferred. However, in speech compression schemes, it is more important to ensure that compression schemes retain the integrity of the speech. For large amount of exchange and transmission of audio information through internet and wireless systems, efficient (i.e., low bit rate) audio coding algorithms need to be devised. The basic task of high-quality audio coding system is to compress the digital audio data in a way the compression is as efficient as possible and the reconstructed audio sounds as close as possible to the original audio before compression.

This chapter highlights the state of the art for digital compression of speech and audio signals. The scope is limited to surveying the most important and prevailing methods, approaches, and activities of current interest without attempting to give a tutorial presentation of specific algorithms or a historical perspective of the evolution of speech coding methods. No attempt is made to offer a complete review of the numerous contributions that have been made in recent years. Nevertheless, the major ideas and trends are covered here and attention is focused on those contributions that have had the most impact on current technology. Hence, efforts have been made to present the comparative study of various types of source coding techniques such as:

1. Linear Predictive Coding (LPC)
2. Code Excited Linear Predictive (CELP) coding
3. Sub-band coding
4. Transform coding
    a. Fast Fourier Transform (FFT)
    b. Discrete Cosine Transform (DCT)
    c. Continuous Wavelet Transform (CWT)
    d. Discrete Wavelet Transform (DWT)
    e. Variance Fractal Compression (VFC)

Lukasaik et al. proposed a low level audio descriptor for psychoacoustic noise. It removes the compression noise than any other descriptor. Here, matching the descriptors' structure to the compression scheme is essential.

Akyol and Rose proposed Karhunen Loeve Transform to minimize mean square error distortion with optimality in transform.

Nanjundaswamy and Rose proposed Cascaded Long-Term Prediction filter for polyphonic signal. While LTP works well for single periodic component, which is not the case with general audio signal, as it consists of multiple periodic signals. LTP is effective for long period stationary signals.

3.2 Linear Predictive Coding

LPC is one of the most powerful and useful speech analysis techniques for encoding good quality speech at a low bit rate. It provides extremely accurate estimates of speech parameters and is relatively efficient for computation. LPC starts with the assumption that the speech signal is produced by a buzzer at the end of a tube. The glottis (space between the vocal cords) produces the buzz, which is characterized by its intensity (loudness) and frequency (pitch). The vocal tract (throat and mouth) forms the tube, which is characterized by its resonances, called as formants. LPC analyzes the speech signal by estimating formants, removing their effects from the speech signal, and estimating the speech intensity and frequency of the remaining buzz. The process of removing the formants is called inverse filtering, and the remaining signal is called residue. Because speech signal vary with time, this process is done on short chunks of speech signal, called as frames. The basic problem of LPC system is to determine the formants from the speech signal.

The solution used is the difference equation, which expresses each sample of signal as a linear combination of previous sample. Such equation is called as a linear prediction, and hence the name given is Linear Prediction Coding.

Linear prediction is an integral part of many modern speech coding systems and is commonly used to estimate the autoregressive (AR) filter parameters. It describes spectral envelope of speech segment. The prediction coefficient is the minimized difference between observed signal and the predicted signal, but the minimization criterion is not optimal in many cases. For example, if the excitation is not Gaussian, the coefficients are found for short-term and long-term signals in two different steps and hence suboptimal solution is obtained. However, the common LP analysis tries to cancel the pitch harmonics by putting some of the poles very close to the unit circle.

The coefficients of the difference equation are prediction coefficients. Prediction coefficients represent the formants, so the LPC system needs to estimate these coefficients. The estimate is done by minimizing the mean square error between predicted signal and actual signal. This is the main problem in principle. The tube must not have any side branches in order to work (side branches introduces more zeros and results in more complex equations). This problem is observed in nasal sound. Another problem in LPC is that any inaccuracy in the estimation of formants leaves more speech information in residue without which quality sound could not be reconstructed. To avoid this, the residue signal may be sent with the compressed one. Unfortunately, the faithful compression will not be achieved. To avoid this problem, LPC is modified as Code Excited Linear Predictive (CELP) coding.

3.3 Code Excited Linear Predictive Coding

It uses codebook, a table of typical residue signals set by the system designers [10]. In operation, the analyzer compares residue to all the entries in the code book, chooses the closest matching entry, and just sends the code for that entry. The synthesizer receives this code, retrieves the corresponding residue from the codebook, and uses that to excite the formant filter. This is called as Code Excited Linear Predictive (CELP) coding. For CELP to work well, the code book must be big enough to include all the various kinds of residues. However, if the code book is too big, it will be time consuming to search through and will require large codes to specify the desired residue. The biggest problem is that such a system would require a different code for every frequency of the source (pitch of the voice), which would make the code book extremely large.

CELP coders are capable of producing good quality speech at around 4.8 Kbps. Below this, they suffer from distortion introduced by coarse quantization of model parameters due to limited number of bits. Also, the frame-by-frame analysis coupled with the high processing demands introduces delay that can degrade quality of conversation and introduce difficulties in related speech processing components, such as echo cancelling.

Most recently, MPEG-4 CELP (2 Kbps in narrowband version and 4 Kbps in Wideband) and MPEG-4 BSAC (Bit Slice Arithmetic Coding) coder (up to 1 Kbps) have been introduced. In this, a core layer produces the lowest bit rate and provides the minimum information to obtain a basic quality for the decoded signal and several enhancement layers within contain additional information to improve quality; however, still the coding efficiency is not comparable.

3.4 Sub-Band Coding

Audio and voice signals have highly nonstationary characteristics. Therefore, attempts to compress such signals typically rely on block processing, where blocks of input samples are first decomposed by a filter bank into sub-band signals. These sub-band signals are in turn analyzed to extract the time-varying signal parameters, which are then input to a coding algorithm. Finally, the coding algorithm transmits a compressed version of the input sample.

We can optimize a given sub-band coding algorithm for nonstationary signals by introducing time-varying filter banks, where for each block of input samples, we alter the structure of the decomposition filter bank such that a particular coding criterion is optimized. The time-varying filter bank design methods employ boundary and entry/exit filters to achieve independent block processing where samples from adjacent blocks do not affect each other. Boundary or entry/exit filters exhibits following characteristics.

- The number of boundary filters and the number of corresponding switching operations increase with the filter length and the depth of the filter bank. These numbers at the synthesis bank are doubled. This creates significant challenges to the design and implementation of computationally efficient time-varying filter banks.

- Individual sub-band signals at the output of time-varying filter bank exhibit severe block effects in the form of abrupt level changes at block boundaries. This reduces the sub-band coding efficiency. Thus, the traditional design techniques do not yield suitable adaptive structure.

Ramya and Sathyamoorthy proposed a speech coder based on sub-band coding at 2.4 Kbps. At the receiver, Mel-Frequency Cepstral Coefficient (MFCC) technique is used to extract the speech parameters along with vector quantization and Huffman coding.

3.5 Transform Coding

Transform coding is the type of data compression for natural data like audio signal or photographic images. In this, the knowledge of the application is used to choose the information to be discarded, thereby lowering its bandwidth. The remaining information can then be compressed via variety of methods. Fast Fourier Transform (FFT), DCT, CWT, DWT, and DPWT are the types of transform coding used for data transformation into another mathematical domain for suitable compression.

3.5.1 Fourier Transform

It is one of the methods for signal and image compression. FT decomposes a signal defined on infinite time interval into a λ-frequency component where λ can be real or complex number. FT is actually a continuous form of Fourier series. FT is defined for a continuous time signal x (t) as,

$$x(f) = \int_{-\infty}^{\infty} x(t) \cdot e^{-i\omega t} \cdot dt \tag{3.1}$$

The above equation is called analysis equation. It represents the given signal in different form, as a function of frequency. The original signal is a function of time, whereas after the transformation, the same signal is represented as a function of frequency. Consider the following two different signals:

$$x1(t) = \sin(2\pi * 100 * t), \text{ for } 0 \le t < 0.1 \sec$$
$$= \sin(2\pi * 500 * t), \text{ for } 0.1 \le t < 0.2 \sec \tag{3.2}$$

$$x2(t) = \sin(2\pi * 500 * t), \text{ for } 0 \le t < 0.1 \sec$$
$$\sin(2\pi * 100 * t), \text{ for } 0.1 \le t < 0.2 \sec \tag{3.3}$$

A plot of these signals is shown in Figures 3.1 and 3.2.

The above example demonstrates the drawback in Fourier analysis of signals. It shows that the FT is unable to distinguish between two different signals. The two signals have same frequency components, but at different times. In general, FT is not suitable for the analysis of a class of signals called nonstationary signals. This problem can be solved by CWT. In FT, all data are smoothened by removing all spike and anomaly; on the other hand this is not the case with DWT.

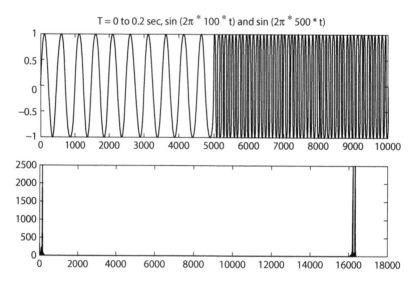

FIGURE 3.1
Signal *x*1(*t*) and its FFT. (From Gunjal, S.D. and Raut, R.D. *Inter. J. Comp. Tech. App.*, 3, 1335–1342, 2012.)

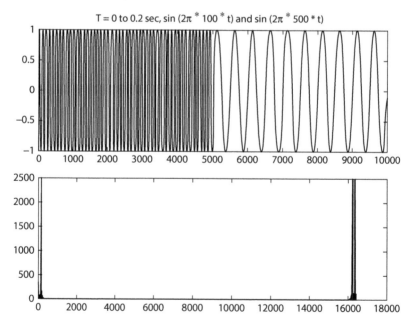

FIGURE 3.2
Signal *x*2(*t*) and its FFT. (From Gunjal, S.D. and Raut, R.D. *Inter. J. Comp. Tech. App.*, 3, 1335–1342, 2012.)

3.5.2 Wavelet Transform

The wavelet transform belongs to the family of filter banks. It consists of a low pass filter followed by decimation of Factor 2. For an N sample input frame, two N/2 sample output frames are called approximation and detail, with bandwidth about half of input signals bandwidth. The low-frequency band is further filtered to provide details at lower resolution levels. The details are not further filtered. This method gives good frequency selectivity at the cost of temporal resolution. However, for high-frequency area, the poor frequency selectivity is not acceptable in speech/audio coding.

Wavelet is a waveform of effectively limited duration that has zero average value. It is the method of compression where it is the ability to describe any type of signal both in time and frequency domains.

Consider a real or complex valued continuous time function ψ (t) with following properties:

1. The function integrates to zero, i.e.

$$\int_{-\infty}^{\infty} \psi(t) dt = 0 \qquad (3.4)$$

2. It is square integral or equivalently has finite energy.

$$\int_{-\infty}^{\infty} |\psi(t)|^2 dt < 0 \qquad (3.5)$$

The function is called mother wavelet if it satisfies Equations 3.4 and 3.5. Fourier analysis consists of breaking up a signal into sine waves of various frequencies. Similarly, wavelet analysis is the breaking up of a signal into shifted and scaled versions of the original (or mother) wavelet.

The most important difference between FT and WT is that individual wavelet functions are localized in space. In contrast, Fourier sine and cosine functions are nonlocal and active for all time t. This in turn results in a number of useful applications such as data compression, detecting features in images, and de-noising signals. The wavelet transform of a signal $f(t)$ is the family C (a, b), given by the analysis Equation 3.6.

$$C(a, b) = \int_{-\infty}^{\infty} f(t) \frac{1}{\sqrt{|a|}} \psi\left(\frac{t-b}{a}\right) dt \qquad (3.6)$$

It depends upon two indices a and b. From an intuitive point of view, the wavelet decomposition consists of calculating a "resemblance index" between the signal and the wavelet located at point b and of scale a. If the index is large, the resemblance is strong and vice versa. The index C (a, b) is called as coefficient.

Let us consider the same example of two signals discussed earlier, demonstrated for FT and show how wavelet analysis distinguishes between the two different signals and also gives their frequency content.

Figures 3.3 and 3.4 show the signals along with their wavelet scalograms. Note the scalograms of these two signals are entirely different, enabling the wavelet transform to distinguish between these two signals.

There are two types of wavelet transforms

1. *CWT* (discussed above) and
2. *DWT*

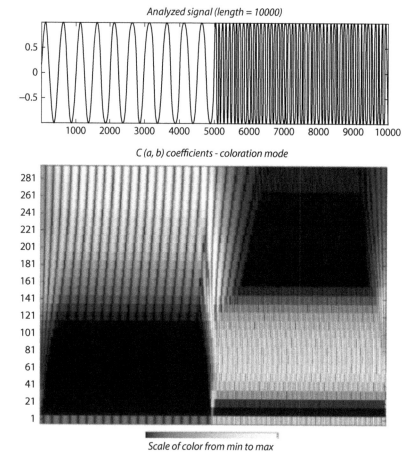

FIGURE 3.3
Signal $x1$ (t) and its scalogram. (From Gunjal, S.D. and Raut, R.D. *Inter. J. Comp. Tech. App.*, 3, 1335–1342, 2012.)

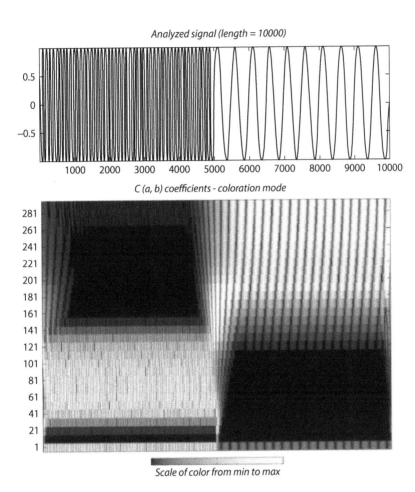

FIGURE 3.4
Signal $x2$ (t) and its scalogram. (From Gunjal, S.D. and Raut, R.D. *Inter. J. Comp. Tech. App.*, 3, 1335–1342, 2012.)

The main idea of wavelet transform is same in both of these transforms. However, they differ in the way the transform is being carried out. The period during the standardization of FFT- and DCT-based compression technology (VFC) has also advanced. However, their compression methodology has not matched the rate-distortion performance of wavelet-based codec.

Wavelet transforms have two additional advantages over DCT that are important for coefficient compression. The first is the multiresolution representation of the signal by wavelet decomposition that greatly facilitates subband coding. The second advantage of wavelet transform is that it reaches a good compromise between frequency and time resolution of the signal. Wavelet transforms are superior to DCT, as their basis functions offer good frequency resolution in the lower-frequency range; at the same time, they yield good time resolution at a higher-frequency range. Noise is extraneous information in a signal that can be filtered out via the computation of averaging and detailing coefficients in the wavelet transform. Wavelet transform is able to detect and localize disturbances. Also, there are several thresholding techniques for wavelet transform coefficients like absolute maximum value, hard threshold, soft threshold, Garrote thresholding, firm thresholding, global threshold, and level-dependent threshold. All the data contents/characteristics are recaptured by multiresolution analysis.

Multiwavelet decomposition is discussed for audio compression. The performance of different types of wavelets for composing the transient audio signal is also discussed. The wavelets that were investigated for this purpose are Daubechies family of wavelets, called wavelet packet and multi-wavelets. One of the main challenges to the application of multiwavelets is the problem of multiwavelet initialization (prefiltering). In the case of scalar wavelets, the given signal data are usually assumed to be the scaling coefficients that are sampled at a certain resolution; hence, multiresolution decomposition can directly be applied on the given signal. Unfortunately, the same technique cannot be applied directly in the multiwavelet setting. Some preprocessing has to be performed on the input signal prior to multiwavelet decomposition. Then, it represents a promising substitute for scalar wavelets in audio compression.

3.5.3 Discrete Wavelet Transform

This is based on sub-band coding and found to yield a fast computation of wavelet transform. It is easy to implement and reduce the computation time and resources required. In CWT, the signals are analyzed using a set of basic functions that relate to each other by simple scaling and transition. In case of DWT, the time scale representation of the digital signal is obtained using digital filtering techniques. The signal to be analyzed is passed through filters with different cutoff frequencies at different scales.

In DWT, a signal can be analyzed by passing it through an analysis filter bank followed by a decimation operation. This analysis filter bank, which consists of a low pass and a high pass filter at each decomposition stage, is commonly used in image compression. When a signal passes through these filters, it is split in two bands. The low pass filter corresponds to an averaging operation and extracts the coarse information of the signal. The high pass filter corresponds to a differencing operation and extracts the detail information. The output of the filtering operation is then decimated by two. Wavelets can be realized by iteration of filters with rescaling as shown in Figure 3.5. This is called as Mallat algorithm or Mallat-tree decomposition.

Here, the signal is denoted by $x(n)$, where n is an integer. The low pass filter is denoted by G_o and high pass filter is denoted by H_o. At each level, the high pass filter produces detailed information $d(n)$, while the low pass

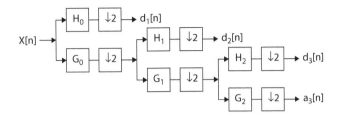

FIGURE 3.5
Three-level wavelet decomposition tree. (From Gunjal, S.D. and Raut, R.D. *Inter. J. Comp. Tech. App.*, 3, 1335–1342, 2012.)

filter produces coarse approximations *a(n)*. At each decomposition level, the half band filters produce signals spanning only half the frequency band. This doubles the frequency resolution as the uncertainty in frequency is reduced by half. In accordance with Nyquist's rule, if the original signal has a highest frequency of ω, which requires a sampling frequency of 200 radians, then it now has a highest of ω/2 radians. It can now be sampled at a frequency of ω radians, thus discarding half the samples with no loss of information. This decimation by 2 halves the time resolution, as the entire signal is now represented by only half the number of samples. Thus, while the half band low pass filtering removes half of the frequencies and thus halves the resolution, the decimation by 2 doubles the scale. The filtering and decimation processes are continued until the desired level is reached. The DWT of the original signal is obtained by coefficients concatenation, that is, starting from the last level of decomposition.

For DWT, various wavelet filters such as Harr (2 filters) and Daubechies (up to 10 filters) are used. All the numerical results were done by using MATLAB programming. Najih et al. discussed the use of wavelet transform for speech compression. Speech compression is the process of converting human speech signals into efficient encoded representation that can be decoded back to produce a close approximation of the original signals. The input signal used is 8 KHz, 8 bit speech. Based on peak signal-to-noise ratio (PSNR), signal-to-noise ratio (SNR), and compression ratio, they concluded that D10 wavelet filter gives higher SNR and better speech quality with compression ratio 4.31 times and reduced bit rate.

Elaydi et al used wavelets to compress speech signal. They used spoken English speech to be analyzed to D20 wavelet filter. With this, compression ratio went higher and SNR was decreased.

Agbiny proposed compression using Battle–Lemarie wavelet, Haar, and Daubechies (up to 20 filters). The analysis was done by voices and unvoiced speech, and the results shows that Battle–Lemarie is the best, while other filters are almost comparable except Haar. Karam and Saad analyzed the effect of different compression schemes on speech signal. They used D4, D8, D10, and D20, and their input is Arabic speech signal. Based on SNR and PSNR, they found that, using smooth wavelets like D10, the percentage of truncated coefficients decreased and give better SNR. For unsmooth wavelet, it gives better compression ratio but low SNR.

Khalifa et al. proposed audio compression by using wavelet transform (up to 10 filters). Transparent coding of audio and speech signals is at the lowest possible data rate. Based on the numerical results, D10 is the best wavelet filter with lowest SNR and highest CR (1.88).

3.5.4 Discrete Wavelet Packet Transform

The wavelet packet method is a generalization of wavelet decomposition that offers a richer range of possibilities for signal analysis. It is used to obtain the critical bands of human auditory system. In wavelet analysis, a signal is split into an approximation and a detail coefficient. The approximation coefficient is then itself split into second level approximation coefficients and detail coefficients and the process is repeated. In wavelet packet analysis, the details as well as the approximations can be split. It gives more than two $2n-1$ different ways to encode the signal.

Figure 3.6 shows the level 3 decomposition using wavelet packet transform. In this case, the entropy-based criterion is used to select the most suitable decomposition of a given signal. This means that we look at each node of decomposition tree and quantity for the information to be gained by performing each split.

The wavelets have several families. The most important wavelets families are Harr, Daubechies, symlet, coiflet, Biorthogonal, reverse Biorthogonal, Meyers, Discrete Approximation of Meyer wavelets, Gaussian, complex, Gaussian, Mexican Hat, Morlet complex Morlet, Ballet Lamarie, and Shannon. Out of these wavelet families, Haar, Daubechies wavelets, symlet, coiflet wavelet families are the most important. For compression purposes, the best cost functional to be minimized is the number of coefficients

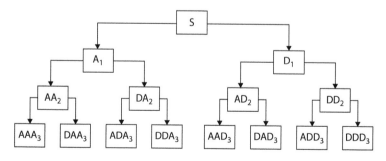

FIGURE 3.6
Level 3 decomposition using wavelet packet transform. (From Gunjal, S.D. and Raut, R.D. *Inter. J. Comp. Tech. App.*, 3, 1335–1342, 2012.)

superior to a certain threshold. This is the reason that the transform based on wavelet packets must be used in compression applications. Use of psychoacoustic model is also suggested in to achieve perceptually transparent compression of high-quality audio signals at about 45 Kbps. The filter bank structure adapts according to psychoacoustic criteria and according to computational complexity that is available at the decoder.

3.6 Variance Fractal Compression

The fractal compression is based on fractal system's ability to approximate discontinuous functions, where audio signals exhibit greater smoothness. It has gained wide popularity due to its inherent features and efficiency in compressing data in medical field. The fractal dimension values indicate the complexity of pattern in terms of morphology, entropy, spectra, or variance. The VFD analysis does not create window artifacts in the Fourier sense, which is often introduced in fractal spectral analysis. Therefore, the VFD is an excellent tool for investigating time series signals by calculating the variance fractal dimensions. This technique can also be integrated into the framework of conventional compression techniques such as vector quantization and transform methods. This direction may also lead to success in applying fractal audio coding in practice.

3.7 Psychoacoustic Model and Daubechies Wavelets for Enhanced Speech Coder Performance

To meet the heavy demands of mobile and Internet users with high fidelity and accuracy, speech compression has become a prime concern for present and future communications. In this regard, speech coding with good quality reproducibility contributes much. Speech compression is one of the most important signal processing steps that can reduce the requirements for transmission bandwidth, storage, and processing overhead time. Many algorithms, such as lossy or lossless algorithms, were proposed earlier for speech signal compression. Ours is a lossy algorithm in which part of the signal information is lost; nonetheless, the loss does not impinge on the reproducibility of the signal. The algorithm offers good compression as well as a good quality speech signal.

In lossy compression, information redundancy is reduced by means of coding. The only information that is necessary to reproduce the signal is kept to ensure good playback quality parametric coders, and waveform coders have also been proposed for compression of the speech signal. A parametric coder, such as the Code Excited Linear Predictive coder, deals with the parameters that characterize filter behavior to encode data, and then the data are used by the decoder for speech synthesis. In contrast, waveform coders attempt to replicate most accurately the waveform of the original signal by exploiting the correlation between the selected filter bank and the signal components in the transform domain.

A transformation is applied to create a spectral representation of the input signal from the corresponding bank of sampled band pass filters. It analyzes the information from the input signal in terms of fewer coefficients to which processing steps such as thresholding and quantization can be applied for further performance improvement. The encoder makes use of the DWT as the main coder, supported by the traditional psychoacoustic model to enhance the compression factor (CF), SNR, and energy content of the speech signal. DWT is an effective mathematical tool for compression of nonstationary signals, as it is a time–frequency localization-based transform. The time–frequency characteristics of a wavelet filter bank match speech signal characteristics very well.

In this chapter, a speech coder based on the DWT using the Daubechies wavelet family is discussed. The Daubechies wavelet family is selected because of its smooth and compactly supported orthogonal low-pass and high-pass FIR filter banks.

Speech compression techniques based on the traditional psychoacoustic model have been proposed by many researchers. The model discussed in this chapter uses the DWT supported by the same psychoacoustic model for speech compression. The model processes equal partitions of the total bandwidth spectrum of audio signal frequencies in order to reduce redundancy by filtering out the tones and noise maskers in the speech signal. To achieve this uniformity, filter banks are used for efficient computation, for selection of appropriate threshold levels, and for better compression of DWT coefficients. A Daubechies wavelet filter bank is nonlinear and asymmetric. It is equivalent to a cochlear filter in the human hearing system. The similarity between the Daubechies filter bank and the hearing system is the basis for development of a novel speech coder.

A block diagram of speech coder deliberated in this chapter is given in Figure 3.7.

It consists of two important blocks:

1. Psychoacoustic model
2. DWT in the form of Discrete Wavelet Decomposition and thresholding

The psychoacoustic model is an important component of the encoder. It is based on research on human perception properties. The two main properties that characterize the psychoacoustic model are the absolute hearing

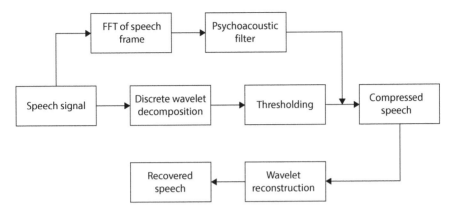

FIGURE 3.7
Block diagram of speech coder. (From Gunjal, S.D. and Raut, R.D. *Inter. J. Tech.*, 2, 190–197, 2015.)

threshold and auditory masking. Even though the human hearing frequency range is 20 Hz to 20 KHz, all frequencies are not heard in the same manner. A person can hear lower frequencies more accurately than higher frequencies. It indicates that our hearing system has a better ability to detect differences in pitch at lower frequencies. In addition, a signal whose frequency component lacks the power specified as the absolute threshold of hearing (ATH) can be removed. A very small frequency difference makes a low power signal inaudible to the listener; hence, the signal needs a power boost greater than the frequency of the masker tone. If the signal is constant except for a sharp peak, then it is considered as a tone; otherwise, it is a noise signal. Noise components that are detected in the critical band are added together to get the resulting noise component. The result of the noise components is used for passing/testing the threshold of the transformed signal. The individual noise masker threshold can be represented by the following equation:

For tones:

$$P_{nm(j)} - 0.27Z_{(j)} + SF_{(i,j)} - 6.025 \tag{3.7}$$

For noise:

$$T_{nm}(i,j) = P_{nm(j)} - 0.175Z_{(j)} + SF(i,j) - 2.025$$

where $SF_{(i,j)}$ is a low level masking noise occurring in the tails of the basilar excitation pattern, and $P_{nm(j)}$ denotes the SPL of the noise masker in frequency bin j. $Z_{(j)}$ represents the bark frequency of bin j. For multiple tones and noise, the overall effect is additive. The psychoacoustic model steps are

Perform FFT analysis:

1. Calculate the energy in each frame.
2. Convolve the energy with the spreading function.
3. Calculate the tonality index (0–1) to separate tones.
4. Calculate the energy threshold for every frame.
5. Compare the value with the absolute threshold of hearing and consider the highest value as the energy threshold of the speech signal.

To overcome the mismatch between uniform filter banks and spectral decomposition of the cochlea, a nonuniform cochlear filter bank is used for the psychoacoustic model. To overcome the above-mentioned problem, traditional psychoacoustic model and the Daubechies wavelet family are used. This is the only nonuniform FIR filter bank that combines perfect signal reconstruction with energy conservation and removes the aliasing effect. In DWT analysis, filters are used with different cutoff frequencies at different scales. DWT offers a compact representation of the signal in the time and frequency domains along with efficient computation in the form of Equations 3.8.

$$d_{jk} = \int x(t)dt = 2^{\frac{j}{2}} \int x(t) \varnothing_{jk} \left(2^j \, t - k \right) dt$$

(3.8)

$$\varnothing_{jk}(t) = 2^{\frac{j}{2}} \varnothing_{jk} \left(2^{-j} \, t - k \right), \quad j, k \in z$$

where d_{jk} is the wavelet coefficient and $x(t)$ is the time signal. The number of vanishing moments is related to the smoothness and flat frequency response of wavelet filters. A large number of vanishing moments produces a more compact signal. A vanishing moment must satisfy the condition given in Equation 3.9.

$$\int x^k \, \psi(x)dx = 0 \quad \text{for, } 0 \leq k \leq K$$

(3.9)

where K is the degree of the polynomial (0 to K). However, the length of filter increases with the number of vanishing moments at the cost of complexity and time for computation. To process a speech signal, a decomposition level up to five is sufficient. Then, the decomposed speech signal is ready for thresholding. Level-dependent thresholding and soft thresholding are applied to the decomposed signal. These types of thresholdings help to modify the CF according to each application's needs. The threshold value is determined using the Birge–Massart strategy, which is well suited for speech signals.

It processes the detail coefficients without disturbing the approximation coefficients. The number of detail coefficients is given by

$$n = M/(j+2-i)^{\alpha} \tag{3.10}$$

where level i starts from 1 to j, α is the compression parameter, and M depends upon the number of approximation coefficients. This thresholding approach provides maximum absolute coefficients at each level. After hard thresholding, the transform vector needs to be compressed further. Here, soft thresholding is used to achieve a good value for the CF. This method has produced improvement in the CF as compared to other methods.

The performance parameters, that is, the CF and SNR of the coder for all five levels, are given in Table 3.1.

The dependencies of CR, SNR, and the decomposition level is shown in Figures 3.8 and 3.9. As per Equation 3.9, the number of vanishing moments is determined by the length of the filter. An increase in vanishing moments naturally supports the increase in CR. However, instead of increasing the complexity of filter design, one can achieve the same end by increasing the number of decomposition levels. From Table 3.1, it can deduced that the compression ratio increases as the level number increases and increase in number of Daubechies filter bank, that is, DB02, DB04, and so on. Nonetheless, further increase in the CR value is limited by the SNR value.

The number of vanishing moments is increased by means of an increase in the decomposition level and in the filter banks. Hence, at a higher power level, the signal part that is essential for the accurate reproduction of the speech signal may get added into the vanishing moment. This causes deterioration of the signal quality in the compressed signal against the noise part. Therefore, the SNR value is higher at the minimum level. There is a slight increase in SNR with the Daubechies filter bank, as shown in Figure 3.9.

TABLE 3.1

Performance Parameters in the Coder

Level	Parameter	DB 02	DB 04	DB 06	DB 08	DB 10
1	CR	1.97	1.98	1.99	1.99	1.99
	SNR	20.38	20.99	21.48	21.50	21.51
2	CR	3.77	3.80	3.83	3.85	3.85
	SNR	17.22	18.20	18.65	18.65	18.70
3	CR	6.68	6.90	6.98	6.99	7.02
	SNR	13.56	14.36	14.87	14.98	15.60
4	CR	7.58	7.71	7.64	7.72	7.73
	SNR	12.07	12.63	13.07	13.07	13.45
5	CR	7.60	8.07	8.53	8.82	9.05
	SNR	11.63	12.07	12.44	12.52	12.70

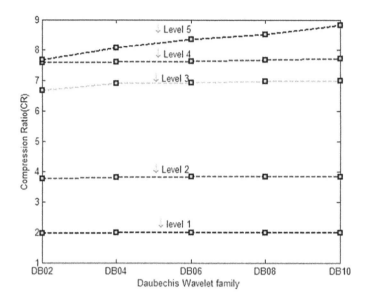

FIGURE 3.8
Daubechies wavelet family vs. CR. (From Gunjal, S.D. and Raut, R.D. *Inter. J. Tech.*, 2: 190–197, 2015.)

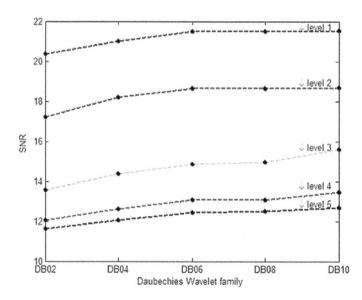

FIGURE 3.9
Daubechies wavelet family vs. SNR. (From Gunjal, S.D. and Raut, R.D. *Inter. J. Tech.*, 2, 190–197, 2015.)

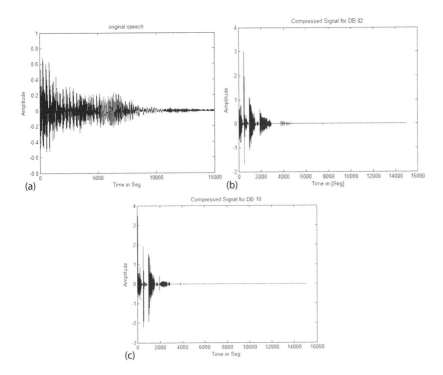

FIGURE 3.10
(a) Original speech signal; (b) compressed speech for DB 02; (c) compressed speech for DB 10. (From Gunjal, S.D. and Raut, R.D. *Inter. J. Tech.*, 2, 190–197, 2015.)

The waveforms for the original speech signal and for the corresponding compressed signal at the fifth level are shown in Figure 3.10. These figures show that increase in the number of levels does not contribute further to listening clarity in the recovered signal.

Figure 3.10 illustrates an example of the compressed speech signal obtained by applying the speech compression technique deliberated in this chapter.

A comparison of the compression ratios obtained from the coder with a classical MPEG1 coder, a coder based on the DWT, and the dynamic gammachirp psychoacoustic model shows that the bit rate of coder based on dynamic gammachirp psychoacoustic model is 372 Kbps and that of the classical MPEG1 coder and DWT coder is 160 Kbps. Moreover, the CR value is even higher than those of the other two models and is found to be 9.05. However, the slower the bit rate, the higher the compression ratio. Thus, it can be concluded that speech compression using the classical psychoacoustic model and the Daubechies Wavelet family is the best.

3.8 Conclusions

This chapter provides a thorough discussion on the parameters such as compression ratio and CF required for the reconstruction of audio signal and also provides a comprehensive review of techniques available in the literature. It is observed that DWT and discrete wavelet packet transform can be used for better results when redundancy with original signal, bit rate, and PSNR are a matter of concern. The use of postfiltering at a suitable bit rate in both transforms improves the quality of the reconstructed signal. It is also suggested that 5 should be the optimum number of wavelet decomposition level. Since high value of wavelet decomposition level will require more computation time, the wavelet packet transform instead of wavelet transform provides the improvement in PSNR of the reconstructed audio signal.

Furthermore, in this chapter, a coder based on traditional psychoacoustic model is discussed. The selection of the Daubechies wavelet family with DWT yielded comparable improvement in the performance parameters with a good quality reconstruction of the speech signal. The CF improves at the cost of the SNR with progressive levels. At levels 3, 4, and 5, the variation in CF and SNR is much more consistent. One can select level 3 for good performance, with a moderate number of filter banks.

Adaptive filter banks can be used in combination with a psychoacoustic model for more effective coding. The coder can further be modified for variable/smaller bit rates for monophonic CD quality audio signals and for mobile communication such as cognitive radio as the energy content of the recovered speech signal is maintained at 98% with slight variations at all levels using the Daubechies filter bank.

References

1. A.M.M. Najih, A.R. Ramli, A. Ibrahim, A.R. Syed. 2003. Speech compression using discrete wavelet transform. *4th IEEE International Conference on Telecommunication Technology Proceedings*.
2. K. Abid, K. Ouni, N. Ellouze. 2010. Audio compression codec using a dynamic gammachirp psychoacoustic model and a DWT multiresolution analysis. *International Journal of Computer Sciences and Engineering*, 2(4): 1340–1354.
3. J.I. Agbinya, 2012. Discrete wavelet transform techniques in speech processing. *IEEE Tencon*, F. Baumgarte, 2002. Improved audio coding using psychoacoustic model based on a cochlear filter bank. *IEEE Transactions on Speech and Audio Processing*, 10(7): 495–503.

4. S. Venkateswaralu, V. Sridhar, R.R.A. Subba, P.K. Satya. 2013. Audio compression using Munich and Cambridge filters for audio coding with Morlet wavelet. *Global Journal of Computer Science and Technology Software and Data Engineering*, 13(5): 1 and D. Pan Motorola, 1995. A tutorial on MPEG/audio compression. *IEEE Transaction on Signal Processing*, 2(2): 60–74.

5. O.O. Khalifa, S.H. Harding, H.A. Aisha-Hashin. 2005. Compression using wavelet transform. *International Journal of Signal Processing*, 2(5): 17–26.

6. S. Krimi, K. Auni, N. Ellouze. 2007. An improved psychoacoustic model for audio coding based on wavelet packet. *IEEE International Conference SETIT 2007*, Hammamet, Tunisia.

7. T. Mourad, C. Barbarnoussi, C. Adnane. 2013. Speech compression based on psychoacoustic model and a general approach for filter bank design using optimization. *International Arab Conference on Information Technology*, ACIT.

8. T. Painter, A. Spanias. 1997. A review of algorithms for perceptual coding of digital audio signal. *IEEE Proceedings on Digital Signal Processing*, 1: 179–208.

9. Y. Shao, C.H. Chang. 2011. Bayesian separation with scarcity promotion in perceptual wavelet domain for speech enhancement and hybrid speech recognition. *IEEE Transaction on System Man and Cybernetics, Part A: System and Humans*, 41(2): 284–293.

10. S.D. Gunjal, R.D. Raut. 2012. Advance source coding techniques for audio/speech signal: A survey. *International Journal for Computer Technology and Applications*, 3(4): 1335–1342.

11. P. Srinivasan, L.H. Jamieson. 1999. High quality audio compression using an adaptive wavelet packet decomposition and psychoacoustic modeling. *IEEE Transactions On Signal Processing*, IEEE.

12. T.A. de Perez, M. li, H. McAllister, H., N.D. Black. 2000. Noise reduction and loudness compression in a wavelet modelling of the auditory system. *IEEE Transaction on Signal Processing*, IEEE.

4

Coding Techniques to Improve Bit Error Rate in Orthogonal Frequency Division Multiplexing System

4.1 Introduction

The telecommunications industry is in the midst of a veritable explosion in wireless technologies [1]. Once exclusively military, satellite and cellular technologies are now commercially driven by ever more demanding consumers, who are ready for seamless communication from their home to their car, to their office, or even for outdoor activities. With this increased demand comes a growing need to transmit information wirelessly, quickly, and accurately. To address this need, communication engineers have combined technologies suitable for high-rate transmission with forward error correction techniques. The latter are particularly important as wireless communications channels are far more hostile as opposed to wire alternatives, and the need for mobility proves especially challenging for reliable communications. Orthogonal frequency division multiplexing (OFDM) is a multicarrier modulation technique in which a single high rate data stream is divided into multiple low rate data streams and is modulated using subcarriers that are orthogonal to each other.

Some of the main advantages of OFDM are its multipath delay spread tolerance and efficient spectral usage by allowing overlapping in the frequency domain. Also, one other significant advantage is that the modulation and demodulation can be done using inverse fast Fourier transmission (IFFT) and fast Fourier transmission (FFT) operations, which are computationally efficient. In a single OFDM transmission, all the subcarriers are synchronized to each other, restricting the transmission to digital modulation schemes. OFDM is symbol based and can be thought of as a large number of low bit rate carriers transmitting in parallel. All these carriers transmitted using synchronized time and frequency, forming a single block of spectrum. This is to ensure that the orthogonal nature of the structure is maintained. Since these multiple carriers form a single OFDM transmission, they are commonly referred to as "subcarriers," with the term "carrier" reserved for describing the radio frequency carrier mixing the signal from base band.

There are several ways of looking at what makes the subcarriers an OFDM signal orthogonal and why this prevents interference between them.

4.2 OFDM System

OFDM is nowadays widely used for achieving high data rates as well as combating multipath fading in wireless communications. In this multicarrier modulation scheme, data are transmitted by dividing a single wideband stream into several smaller or narrowband parallel bit streams. Each narrowband stream is modulated onto an individual carrier. The narrowband channels are orthogonal vis-à-vis each other and transmitted simultaneously. In doing so, the symbol duration is increased proportionately, which reduces the effects of intersymbol interference (ISI) induced by multipath Rayleigh-faded environments. The spectra of the subcarriers overlap each other, making OFDM more spectral efficient as opposed to conventional multicarrier communication schemes.

4.2.1 OFDM Message

The OFDM message is generated in the complex baseband. Each symbol is modulated onto the corresponding subcarrier using variants of phase shift keying (PSK) or different forms of quadrature amplitude modulation (QAM). The data symbols are converted from serial to parallel before data transmission. The frequency spacing between adjacent subcarriers is $N\pi2$, where N is the number of subcarriers. This can be achieved by using the inverse discrete Fourier transform (IDFT), easily implemented as the inverse fast Fourier transform (IFFT) operation. As a result, the OFDM symbol generated for an N-subcarrier system translates into N samples, with the ith sample being

$$X_i = \sum_{n=0}^{N-1} C_n^{j2\pi in/N}, \ 0 \leq i \leq N-1 \tag{4.1}$$

At the receiver, the OFDM message goes through the exact opposite operation in the discrete Fourier transform (DFT) to take the corrupted symbols from a time domain form into the frequency domain. In practice, the baseband OFDM receiver performs the fast Fourier transform (FFT) of the receive message to recover the information that was originally sent (Figure 4.1).

4.2.2 Interference

In a multipath environment, different versions of the transmitted symbol reach the receiver at different times. This is due to the fact that different propagation paths exist between transmitter and receiver. As a result,

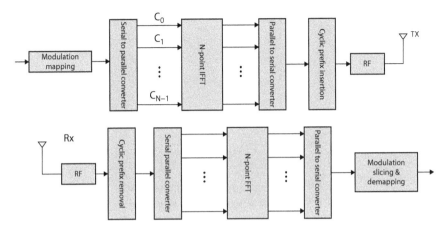

FIGURE 4.1
Basic OFDM architecture. (From Agrawal, D.G. et al., *Int. J. Comput. Technol. Electron. Eng.*, 2, 94–102, 2011.)

the time dispersion stretches a particular received symbol into the one following it. This symbol overlap is called intersymbol interference, or ISI. It also is a major factor in timing offset. One other form of interference is intercarrier interference or ICI. In OFDM, successful demodulation depends on maintaining orthogonality between the carriers. A specific subcarrier N is demodulated at its spectral peak, meaning that all the other carriers must have a corresponding zero spectra at the Nth center frequency (frequency domain perspective). Frequency offsets lead to this criterion not being met. This condition can seriously hinder the performance of the OFDM system. Figure 4.2 shows that when the decision is not taken at the correct center frequency (i.e., peak) of carrier considered, adjacent carriers factor in the decision making, thus reducing the performance of the system.

4.2.3 The Cyclic Prefix

OFDM demodulation must be synchronized both in the time domain and in the frequency domain. Engineers have found a way to ensure that goal by adding a guard time in the form of a cyclic prefix (CP) to each OFDM symbol. The CP consists in duplicating of the end samples of the OFDM message relocated at the beginning of the OFDM symbol. This increases the length T_{sym} of the transmit message without altering its frequency spectrum. $T_{sym} = CP + T_{data} N$, where T data is the duration of one data symbol and N is the number of carriers. The receiver is set to demodulate over a complete OFDM symbol period, which maintains orthogonality. As long as CP is longer than the channel delay spread, τ_{max}, the system will not suffer from ISI. The CP is to be added after the FFT operation at the transmitter and removed

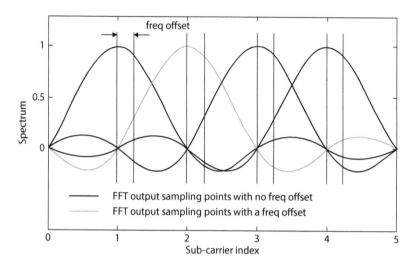

FIGURE 4.2
Effect of frequency offset (maintaining orthogonality). (From Agrawal, D.G. et al., *Int. J. Comput. Technol. Electron. Eng.*, 2, 94–102.)

prior to demodulation. Figure 4.3 shows the deterioration in performance when the CP is closely matched by the delay spread.

The signal constellation is less tightly grouped, which is without doubt a sign of less-than-accurate decoding.

The block diagram of cyclic prefix is shown in Figure 4.3.

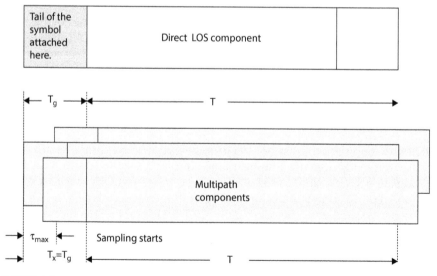

FIGURE 4.3
Cyclic prefix. (From Agrawal, D.G. et al., *Int. J. Comput. Technol. Electron. Eng.*, 2, 94–102.)

4.3 Performance Analysis of Uncoded and Coded OFDM System for Worldwide Interoperability for Microwave Access Networks

The Worldwide Interoperability for Microwave Access (WiMAX) is a rising global wireless system based on IEEE 802.16 standard. WiMAX is an OFDM-based technology that promises high data rate services with wide area coverage with large user densities having a variety of QoS requirements. WiMAX is able provide broadband wireless access up to 30 miles (50 km) for fixed station and 3–10 miles (5–15 km) for mobile stations with high data rates of about 1.5 and 75 Mbps per channel. The WiMAX standard air interface comprises the characterization of both the medium access control and the physical (PHY) layers for the user and the base station, while the access network operability is defined by the WiMAX Forum.

One of the important mechanisms is to combine modulation scheme with forward error correcting (FEC) codes. A high data transmission speed is provided by a high-order modulation scheme, but it causes more susceptibility to interference. FEC code causes redundancy in the data transmission by repetition of some of the data bits, so bits that are missing or in error are corrected at the receiver end. This helps to reduce latency by cutting down the retransmissions. Without the FEC technique, we need whole frames to be retransmitted, which results in latency and lower QoS. There are three basic types of FEC codes: the block codes, the convolutional codes, and the turbo codes. In this section, the performance of the PHY layer of the WiMAX system is analyzed with and without turbo codes.

The function of the WiMAX PHY layer is the actual PHY transfer of data from sender end to receiver end. To gain maximum performance for high data transmission rate (both in fixed and mobile environments) and high spectral efficiency with varied QoS requirements, this supports some PHY mechanisms with different features including an OFDM technique. This is a multicarrier technique that is bandwidth efficient and splits the bandwidth of a system into orthogonal subchannels, where each subchannel occupies a narrow bandwidth and a separate subcarrier is assigned to every subchannel. With the help of guard interval and cyclic prefix, an OFDM system is able to achieve better resistance against multipath fading. OFDM technology offers an efficient means to overcome the challenges of NLOS propagation.

WiMAX supports link adaptation techniques known as adaptive modulation and coding that allow the WiMAX system to adjust the signal modulation scheme depending on the signal-to-noise ratio (SNR) condition of the radio link. Now when radio frequency link quality is high, the highest modulation scheme is selected for use, it gives the system more capacity. Now if a signal is faded, the WiMAX system turns to a lower modulation scheme to preserve

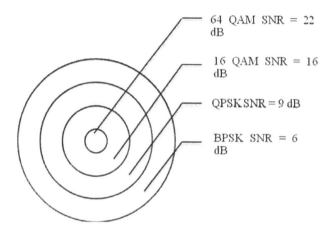

FIGURE 4.4

Relative cell radii for adaptive modulation. (From Nemade, S. et al., *Int. J. Appl. Innov. Eng. Manag.*, 2, 2013.)

the quality of connection and link stability. Because of this feature, the system is able to overcome time-selective fading.

Error correction mechanism has been incorporated into WiMAX mainly to reduce the SNR requirements of system. FEC codes, convolutional encoding, turbo coding, and interleaving algorithms are adapted to identify and correct all errors to improve the throughput of system. These error correction techniques are helpful in recovery of frames which have errors. This considerably helps to improve the bit error rate (BER) performance at related threshold level.

Figure 4.4 shows the relative cell radii for adaptive modulation.

4.4 Turbo Coding in IEEE 802.16e WiMAX

Turbo codes are provided in IEEE 802.16 standard as an optional channel coding scheme. These Turbo codes are fundamentally convolutional codes that are concatenated in series or in parallel manner. These codes are also known as convolutional turbo codes. The WiMAX uses duo binary turbo codes in which a pair of bits is get used in both the regular and interleaved coding iterations for encoding, with a basic recursive encoder of limit length 4 bits. The fundamental WiMAX system model is developed and analyzed as shown in Figure 4.5.

Data bits

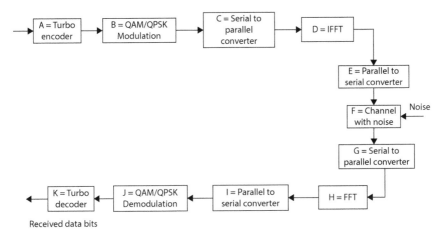

FIGURE 4.5
Simulation model of turbo code OFDM for WiMAX PHY layer model developed in Matlab 2013. (From S. Nemade et al., *Int. J. Innov. Eng. Res. Manag.*, 2, 2013.)

To simulate the effective OFDM transmission scheme, the whole design of system is further divided into three sections: (1) transmitter, (2) receiver, and (3) channel.

The performance of the turbo-coded OFDM is evaluated through the MATLAB simulation model. This simulation follows the procedure as listed below:

1. Generate the information bits randomly.
2. Encode the information bits using a turbo encoder with the specified generator matrix.
3. Use QPSK or different QAM modulation to convert the binary bits, 0 and 1, into complex signals (before these modulation use zero padding).
4. Perform serial-to-parallel conversion.
5. Use IFFT to generate OFDM signals, zero padding is done before IFFT.
6. Use parallel-to-serial convertor to transmit signal serially.
7. Introduce noise to simulate channel errors.

   It is assumed that the signals are transmitted over an AWGN channel. The noise is modeled as a Gaussian random variable with zero mean and variance σ^2. A built-in MATLAB function random is used to generate a sequence of normally distributed random numbers, where randn has zero mean and 1 variance. Thus, the received signal at the decoder is $X' = $ noisy (X), where noisy (X) is the signal corrupted by noise.

8. At the receiver side, perform reverse operations to decode the received sequence.

9. Count the number of erroneous bits by comparing the decoded bit sequence with the original one.

10. Calculate the BER and plot it.

During these simulations, in order to compare the results, the same random messages are generated. For this, radiant function is used that is available in MATLAB.

A comparison of the performance of hard- and soft-decision turbo-coded OFDM systems with 16-Quadrature Amplitude Modulation (16-QAM) and 64 QAM is considered. The various parameters that are used for schematic evaluation are summarized in Table 4.1.

A performance of PHY layer of WiMAX system with turbo coding against different modulation techniques and different channel conditions is measured, and the BER is calculated that forms some curves. The comparison of the performance of WiMAX PHY layer turbo coded over AWGN channel for 16 QAM and 64 QAM systems is shown in Figures 4.6 and 4.7, respectively. The results show that the higher the modulation scheme, the higher is the performance at the high values of SNR, while lower-order schemes perform much better at lower SNRs.

However, the gain should be observed at the expense of longer delays in processing of the signals which results from the increased complexity of the system and can be overcome with improved turbo encoder design.

TABLE 4.1

Parameters and Values

Parameter	Value
Digital Modulation	16 QAM 64 QAM
Turbo Code Rates	1/2
Channel Model	AWGN channel
CP Length	1/8
Channel Bandwidth	5 MHz
FFT Size	256
No. of Users	1

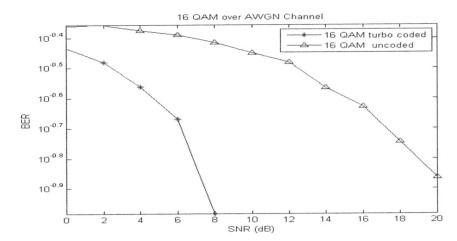

FIGURE 4.6
Performance analysis of 16-QAM uncoded and turbo coded over AWGN channel. (From Nemade, S. et al., *Int. J. Innov. Eng. Res. Manag.*, 2, 2013.)

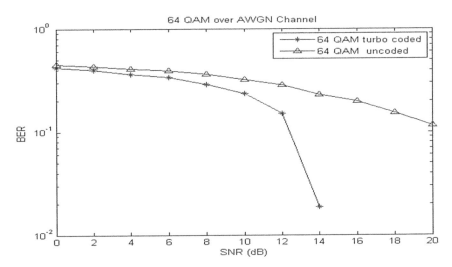

FIGURE 4.7
Performance analysis of 64-QAM uncoded and turbo coded over AWGN channel. (From Nemade, S. et al., *Int. J. Innov. Eng. Res. Manag.*, 2, 2013.)

4.5 Artificial Intelligence Enabled Turbo-Coded OFDM System for Improved BER Performance

The BER performance of any wireless communication system is the primary factor of consideration for the network under evaluation. Turbo-coded OFDM systems have always outperformed other communication standards in terms of delay of communication, overall system efficiency, and flexibility. However, the BER performance of turbo-coded OFDM systems has always been a challenge for network designers due to the fact that turbo encoders vary largely in terms of their primary specifications like code length and code rate among other network parameters. In this section, an artificial intelligence (AI) layer for turbo-coded OFDM systems, which improves the BER performance of the system, is studied. It also reduces the delay of communication by intelligent selection of network specifications. The AI is trained for minimizing the BER of the communication network, but it can be used to target one or many parameters as per the requirements of the network designer.

Wireless communication has always suffered from data packet drops, which occurs due to the fact that all the data packets being communicated go through a wireless channel. This wireless channel introduces inherent noise into the system, thereby corrupting the data transmitted over the channel. These noises can be either additive, reflective, or power absorbing in nature, and they produce a similar effect on the data on the channel. All these noise types affect one parameter in common, which is termed as BER of the system. While a lower BER is expected from a wireless communication system, practical wireless networks usually are incapable of combating this issue. The result is low quality of signals, missing of critical information from the data, and, in some extreme cases, causing network failure.

Many researchers have tried to solve the issue of BER reduction by incorporating different modulation techniques, coding techniques, and even channel modeling for denoising of the received signal. All these techniques are capable of reducing the BER of the wireless communication network to an extent, but all of them perform well under one or two network configurations and may not give desirable network performance when it comes to practical applicability under randomly varying network conditions. Even with so many complications, one particular source coding technique, namely, turbo codes combined with OFDM modulation, provides a high quality of network performance under random and unpredictable network conditions. Turbo codes have three main components when it comes to network design:

- High-performance interleaver
- Convolution coder
- Viterbi decoder with synchronized de-interleaver.

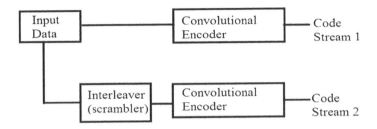

FIGURE 4.8
Turbo code architecture.

The overall turbo codes architecture can be seen in Figure 4.8. Here, the message signal X is first interleaved to form a jumbled-up signal Y, the signal Y has the same bit contents as that of X, but they are randomly swapped in order to create a signal that is an unreadable form of the input signal X. Both the signals X and Y are given to two different convolutional encoders. These convolutional encoders can have different bit lengths, and different code rates for security and code efficiency purposes. Usually, the turbo-coded systems are cascaded with OFDM communication stack, as demonstrated in Figures 4.9 and 4.10. OFDM adds orthogonal channel modulation

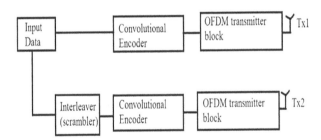

FIGURE 4.9
Turbo OFDM transmitter.

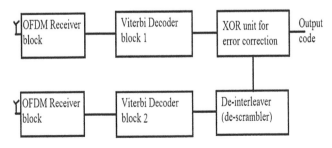

FIGURE 4.10
Turbo OFDM receiver.

to the turbo codes, and thus is useful to reduce the errors introduced during signal propagation over the channel. The OFDM communication stack first converts the bit stream into symbols via digital modulation techniques like BPSK, QPSK, or QAM. These symbols are then given to an IFFT block for frequency division and are then finally analog modulated using analog transmitters. The recovered signal after OFDM demodulation is given to a set of two Viterbi decoders, which are analogous to the convolution encoders used during encoding.

The outputs of the Viterbi decoders are then de-interleaved using a synchronous de-interleaver, which rearranges the bit stream in order to produce a signal that should be similar to the original transmitted bit stream. The difference between the original bit stream and the received bit stream is evaluated in order to evaluate the BER of the wireless communication system. To reduce BER of the turbo-coded OFDM system an AI model is introduced between the turbo coder and the turbo decoder. The AI layer initially trains itself under different channel conditions and tries to reduce the BER by varying the turbo coder/decoder configurations in terms of code length, code rate, channel SNR, and interleaved sequence used; these variations are recorded for minimum BER, and once the communication system is sufficiently trained, it can be used under any channel condition and guarantees a minimum BER as decided by the system designer. The AI system is trained to provide a BER below 0.02, and it achieves that goal after nearly 16–18 generations of training.

Usually, space–time-coded orthogonal frequency division multiplexing (STCOFDM) system is a good communication standard for broadband systems. However, the bit rates offered by this system are limited and not suited for voice and Internet data traffic. Space–time coding (STC) is used in MIMO communications with OFDM systems or other modulation schemes. Spatial and temporal diversity of antennas is best used by the space–time turbo codes [2]. Space–time codes are of two types: block codes and trellis codes. Space–time trellis coding uses convolutional encoders on the transmitter side and Viterbi decoder on the receiver side. Low delay and low BER are the advantages of STC, which makes it useful for high rate wireless applications. Initial STC research efforts focused on narrowband flat-fading channels. STTCs have an exponentially increasing decoding complexity as a function of the diversity level and transmission rate.

Among the existing wireless techniques, OFDM has been most advantageous and gained a lot of popularity among researchers and network designers. Recent wireless protocols, like ultra wideband , advanced source and channel encoding, various smart antenna techniques, space–time codes (STCs), space division multiple access (SDMA), beam forming, and other multiple-input multiple-output (MIMO) wireless architectures offer substantial improvement in performance. Obtaining the maximum possible diversity gain is the primary objective of the space–time block coding (STBC) techniques as well as space–time trellis coding (STTC) schemes [3].

The improvement in performance of second-order transmitters and up to sixth-order receiver diversity can be found in the context of STBC-aided MIMO OFDM. Finally, beam formation removes any interference in the bands of the primary users from that of the secondary receivers, provided that their received signals can be separated angularly. Transmission rate, the transmission range, and the transmission reliability define the three basic parameters for QOS of wireless systems. Usually, the transmission rate is improved by reducing the other two parameters. In contrast, the transmission range can be improved with the performance degradation of a lower transmission rate and reliability, while the transmission reliability may be optimized by reducing the transmission rate and range. But, MIMO-assisted OFDM systems can improve all the three parameters due to the system design proposed by the MIMO OFDM communication architecture.

Real-time implementation of MIMO-based OFDM systems has shown that increased capacity, coverage, and reliability can be obtained practically with the use of the MIMO OFDM architecture [4]. Usually, MIMOs can be combined with any type of wireless communication standard; however, in practice, there is a significant performance enhancement of MIMO-aided OFDM over the non-MIMO OFDM technique. Capacity and coverage is improved with the help of beam forming MIMO OFDM communication systems, and it has already been tested under various extreme channel conditions. Smart antenna techniques, which use strong spatial correlation for processing the received signal by an array of antennas with beam-forming techniques, are able to provide high-directional beam-forming gain and also reduce the interference from other undesired directions under high spatial correlated MIMO channel. The three techniques, MIMO, OFDM, and beam forming, when combined, can have significant improvement in performance as compared to normal nonhybrid system.

These three techniques have similar features that require multiple antenna elements and do not have the contradictive requirement for spacing of antenna elements. Because it is conflictive, the smart antenna works under high spatial correlated MIMO channel, while the spatial diversity technique works under low spatial correlated MIMO channel. Thus, the combination of STC and OFDM is a promising scheme for future wideband multimedia wireless communication systems. However, using multiple antennas causes interference issues between the desired transreceiver pair and the other pairs connected in series with the given transreceiver pair. To minimize the interference from other nonrequired directions, smart antennas are used at the receivers that reject the beams from other directions and enhance the beams from the required directions by using effective beam formation. Turbo-coded OFDM system performance can be further enhanced by application of machine learning techniques. The performance evaluation of the AI-based turbo-coded OFDM system is done under various network conditions.

The AI turbo-coded layer is formed using the following components [5]:

- Turbo encoder/Viterbi decoder
- OFDM transreceiver
- Random code generator
- BER checker
- Feedback device
- Learning component

The turbo encoder and Viterbi decoder perform the standard task of encoding and decoding the input data stream, while the OFDM transreceiver performs wireless communication over varying channel conditions. The AI layer first needs to be trained under different network conditions, and this training would attune the communication parameters of the turbo-coded OFDM system. Training of the AI layer works in the following parts:

- Generation of random turbo configurations
- Evaluation of configuration BER
- Selection of best configuration under a given network condition

The first step is to generate N random turbo configurations, which include variation in code length, code rate, interleaver sequence, and channel SNR. The network is configured with these generated parameters, and the BER of the system is evaluated. The evaluated BER is then compared with the minimum threshold requirement of the BER, and if the generated configuration fulfills the minimum BER requirement, then it is stored for future usage. This process is repeated for "k" iterations, and finally we get a trained AI, which consists of channel configurations that satisfy the minimum BER requirements that are expected from the designed network. The design parameters are defined as follows. Generation of solutions for the turbo-coded system can be done using Equations 4.2–4.4.

$$\text{Code length} = \text{GC} \times \text{Initial code Length} \qquad (4.2)$$

$$\text{Input sequence} = \text{Binary GC of length same as input sequence} \qquad (4.3)$$

$$\text{Channel SNR} = \text{GC} \times \text{Max channel SNR} \qquad (4.4)$$

Interleaved sequence = Numeric GC of length same as input sequence here GC = randomized gold code generator initial code length is kept as 8, and then it is increased from 8 to 512 in powers of 2, this is done in order to find the best code. Figure 4.11 shows the noise variance versus frequency.

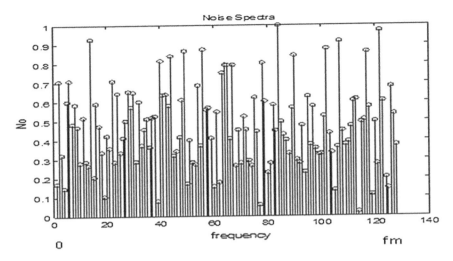

FIGURE 4.11
Noise variance versus frequency.

Three channel models are considered, namely, AWGN, Rayleigh, and Rician. The AWGN channel can be modeled as shown in Equation 4.5

$$p(r) = x(r) + n(r) \tag{4.5}$$

where $p(r)$ is the instantaneous value of the noisy signal, while $x(r)$ is the input signal and $n(r)$ is the random noise signal. The AWGN channel only adds noise to the signal, while the Rayleigh channel fades the signal, which diminishes the real value of the signal, thereby reducing the overall amplitude of the signal. The Rayleigh channel is modeled using Equation 4.6

$$p(r) = \begin{cases} \dfrac{r}{\sigma^2} \exp\left[\dfrac{-r^2}{2\sigma^2}\right], & 0 \le r \le \infty \\ 0, & r < 0 \end{cases} \tag{4.6}$$

here r is the input signal and sigma is the standard deviation of the channel. When the noise is plotted against frequency, then the response looks similar to Figure 4.11. The Rician fading channel model is similar to that of Rayleigh fading, but it adds reflective elements to the model, and can be represented as follows using Equation 4.7.

$$p(r) = \begin{cases} \dfrac{r}{\sigma^2}\exp\left[\dfrac{-r^2}{2\sigma^2}\right]+\varsigma(r), & 0 \le r \le \infty \\ 0, & r < 0 \end{cases} \tag{4.7}$$

In this equation, tau (r) is the instantaneous reflection of the signal "r" caused due to the Rician channel. The analysis shows that the turbo-coded OFDM systems perform best under AWGN channel, and then Rician and Rayleigh channels. This performance can be improved by exhaustively training the system. The training of the AI layer usually takes between 1 min and 4 days depending on the number of iterations, the code length, and the required BER values. For test, 100 iterations are taken with a maximum code length of 512 bits, and the system got trained in 2 min, 32 sec on a Core i5 processor with 16 GB of RAM. Once the training is completed, the AI layer is placed between the transmitter and receiver sections, and it gives signals to both the communication ends. These signals consist of the network configuration to be used, which is selected based on the number of bits to be transmitted, the expected minimum channel SNR, and the channel type. The AI layer selects these parameters intelligently for the network and sets up the transmitter and receiver for communication. As all these combinations have already been tested for the minimum BER requirement, the overall system BER is below the given threshold.

The system is tested under various channel conditions, and various turbo code configurations. Table 4.2 shows the performance in terms of delay and BER of the system. The AI layered approach performs very well under various channel conditions, and the BER response is quite impressive, with the system reducing the BER to below 0.05, with a delay of less than 1 nanosecond (ns) for each of the channels. The simulation parameters used in the design are defined in Table 4.3; these parameters are standard and are generally used by real-time simulation systems for testing the performance of networks. The overall system performance can be depicted by the graphs shown in Figures 4.12 and 4.13. The delay performance of the system starts increasing linearly as number of communications is increased, but it becomes almost constant or saturates around the 0.85–0.95 ns level.

The delay for communication is almost independent of the channel under use. It changes marginally due to the fact that OFDM communication is used, which changes the detection BER with changes in the channel type. BER is generally found to be in the 0.02–0.01 level due to the AI layered communication system, but the system performance under AWGN channel is marginally better than that under Rayleigh, while Rician channel falls a bit behind when compared to AWGN. This is due to the fact that Rician channel distorts the signal to a higher level than AWGN or Rayleigh, so the BER under AI communication takes a little performance hit, but still it is in the 0.01–0.02 range bracket. Figure 4.14 shows experimental results of BER without AI and BER with AI.

TABLE 4.2

BER and Delay under Various Channels and Communications

Channel Type	No. of Communications	Code Length (Bytes)	BER	Mean Delay (ns)
AWGN	10	32	0.02	0.1
AWGN	20	64	0.018	0.3
AWGN	30	128	0.017	0.4
AWGN	40	256	0.0165	0.45
AWGN	50	512	0.0162	0.55
AWGN	100	1024	0.015	0.65
AWGN	500	2048	0.013	0.75
AWGN	1000	4096	0.011	0.88
AWGN	10,000	8192	0.009	0.95
Rayleigh	10	32	0.025	0.15
Rayleigh	20	64	0.023	0.25
Rayleigh	30	128	0.022	0.35
Rayleigh	40	256	0.02	0.5
Rayleigh	50	512	0.019	0.6
Rayleigh	100	1024	0.0187	0.7
Rayleigh	500	2048	0.0184	0.76
Rayleigh	1000	4096	0.0181	0.9
Rayleigh	10,000	8192	0.0176	0.94
Rician	10	32	0.028	0.18
Rician	20	64	0.019	0.28
Rician	30	128	0.0175	0.37
Rician	40	256	0.0164	0.49
Rician	50	512	0.0157	0.56
Rician	100	1024	0.0149	0.67
Rician	500	2048	0.0141	0.71
Rician	1000	4096	0.012	0.82
Rician	10,000	8192	0.01	0.91

TABLE 4.3

Design Parameters

Parameter Name	Value
Channels used	AWGN, Rayleigh, Rician
Turbo coder	Interleaved coder with $k = 1/3$, variable length
Random number polynomial generator	Gold code generator
Modulation used	OFDM with QAM modulation

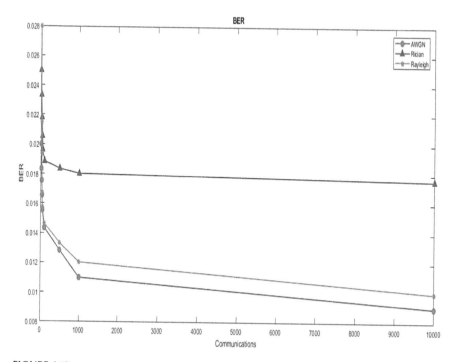

FIGURE 4.12
BER v/s number of communications. (Redrawn, no permission required.)

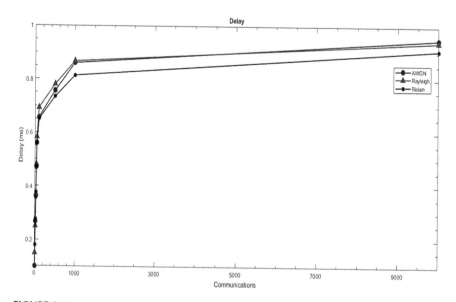

FIGURE 4.13
Delay of detection v/s number of communications. (Redrawn, no permission required.)

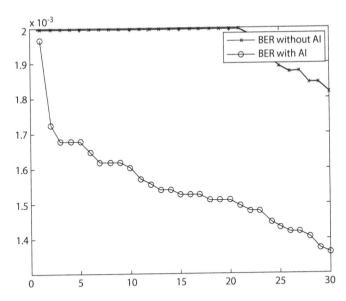

FIGURE 4.14
BER without AI and BER with AI.

Initially without applying AI network, BER is calculated with respect to SNR. For optimization, we had used AI network with deep learning technique. Deep learning AI is trained and can be used in any network condition with any network configuration and will guarantee the minimum BER requirement.

4.6 Conclusions

The chapter begins with an introduction to OFDM system. Next, the performance of IEEE 802.16d/e based WiMAX PHY layer is studied by simulating its different key aspects. The simulation of the optional feature of turbo coding in the PHY layer of the WiMAX system is discussed. The result shows that performance of WiMAX system can be optimized to a lower bit error. The improvement of WiMAX PHY layer performance basically for a single mobile user is discussed.

Furthermore, AI-enabled turbo-coded OFDM system for improved BER performance is further deliberated in this chapter. Low BER and a low delay for network communication are realized. The system architecture ensures that the BER is always below the specified threshold value. The overall system BER is less than 0.02 under different channel scenarios and under varying node communications. As the delay of communication is very less, this network can be used in real-time radio environments, and will be helpful for hardware implementation of the turbo-coded AI layer system.

References

1. D.G. Agrawal, R.K. Paliwal, P. Subramanium, 2011. Effect of turbo coding on OFDM transmission to improve BER. *International Journal of Computer Technology and Electronics Engineering*, 2: 94–102.
2. V. Tarokh, H. Jafarkhani, A.R. Calderbank, 2006. Space-time block coding for wireless communications: Performance results. *IEEE Journal on Selected Areas in Communications*, 17(3): 451–460.
3. V. Tarokh, A.F. Naguib, N. Seshadri, A.R. Calderbank, 1999. Bank, space time codes for high data rate wireless communication: Performance criteria in the presence of channel estimation errors, mobility, and T. multiple paths. *IEEE Transactions on Communications*, 47(2): 199–207.
4. S.M. Alamouti, 1998. Simple transmit diversity technique for wireless communications. *IEEE Journal on Selected Areas in Communications*, 16(8): 1451–1458.
5. P. Subramanium, R.D. Raut, 2018. AI-enabled turbo-coded OFDM system for improved BER performance. *The Journal of Supercomputing*, doi:10.1007/s11227-018-2370-1.

5

Digital Modulation Techniques for Software Defined Radio Applications

5.1 Introduction

Today's wireless networks consist of a large array of mobile equipments. The communication between variety of mobile equipments is regulated by different IEEE standards [1]. The software defined radio (SDR) provides greatest advantage by its reconfigurable front-end capability. IEEE Standard 802.11a boasts of impressive performance. It is able to transmit at the data rates of up to 54 Mbps. The summary of 802.11a Wi-Fi standard is given in Table 5.1.

SDR forum has defined SDR as a radio in which some or all of the physical layer functions are software defined. The software is used to determine the specification of the radio and what it does. If the software within the radio is changed, its performance and function may change. SDR has generic hardware platform to implement modulation and demodulation functions. It also involves filtering, changes in bandwidth, frequency selection, and, in some cases, frequency hopping. These devices include field programmable gate array (FPGA), digital signal processors (DSPs), general-purpose processor (GPP), programmable system on chip (SoC), or other application-specific programmable processors.

For a wireless communication, quality of service is fundamental reference for all network planification. The main quality of service aspects are elaborated as follows:

1. Transmission quality is related with transmitted information fidelity. Information emitted from the sender over communication system must arrive to the receiver without errors, alteration, and losses. The quality global criterion depends on the service types such as legibility in communication, quality and conformity in image transmission, fidelity, purity in musical transmission, and rate error probabilities in data transmission.

2. Other technical factors required to be considered are the total attenuation of liaison of propagation, bandwidth, the comportment

TABLE 5.1

802.11a Wi-Fi Standard Summary

Modulation	Coding Rate (R)	Coded Bits per Sub Carrier (NBP SC)	Coded Bits per OFDM Symbol (NCB PS)	Data Bits per OFDM Symbol (ND BPS)	Data Rate (Mb/s) (20 MHz Channel Spacing)
BPSK	1/2	1	48	24	6
BPSK	3/4	1	48	36	9
QPSK	1/2	2	96	48	12
QPSK	3/4	2	96	72	18
16-QAM	1/2	4	192	96	24
16-QAM	3/4	4	192	144	36
64-QAM	2/3	6	288	192	48
64-QAM	3/4	6	288	216	54

with distortions, perturbations influences with noise, and diaphony. Today's SDR technology is required to handle multiple waveforms, modulation techniques, pulse shaping techniques, and transmit power. The important factors deciding the choice of modulation scheme are as follows:

a. Spectrally efficient modulation, which gives least amount of interference for adjacent channel and neighboring channels.

b. Robust performance in fading multipath fading channels, Doppler frequency.

c. How does the bit error rate (BER) vary with the energy per bit available in the system when white noise present?

d. the cost efficient modulation scheme.

e. easy to implement circuitry, small size, and weight.

The modern wireless communication devices require higher bit rates. Hence, to increase the speed of information transmission, bit rate can be increased by sending more number of bits per symbol, with the help of advanced modulation techniques. The bit rate can be increased by providing larger bandwidths, which gives higher symbol rates resulting in higher bit rates.

5.2 Digital Modulation Techniques

In telecommunication, modulation is the process of conveying a data signal, inside another signal, which is then physically transmitted. Modulator is the device, which modulates the characteristics of a high-frequency carrier wave as per the input data signal. Modulators work on the principle of different modulation techniques, which are mainly divided into two streams [2]:

1. Digital modulation
2. Analog modulation

Digital modulation: Digital modulation is a technique where an analog carrier signal is modulated by a discrete signal. Digital modulation methods can be considered as digital-to-analog conversion, and the corresponding demodulation. It is further categorized in different schemes based on what parameters the characteristic of the carrier wave is modulated.

1. Phase shift keying
2. Amplitude shift keying
3. Frequency shift keying
4. Quadrature amplitude modulation (QAM)
5. Orthogonal frequency division multiplexing (OFDM)

Analog modulation: Analog modulation transfers an analog data signal over an analog band pass channel at a different frequency. Analog modulation schemes aim at transferring a narrowband analog signal over an analog baseband channel as a two-level signal by modulating a pulse wave.

1. Amplitude modulation
2. Phase modulation
3. Frequency modulation
4. Double sideband modulation
5. Single-sideband (SSB) modulation

The different digital modulation schemes such as BPSK, QPSK, and QAM etc. are discussed in this section.

5.2.1 Binary Phase Shift Keying

In binary phase shift keying (BPSK), the phase of a constant amplitude carrier is switched between two values according to the two possible signals m_1 and m_2 corresponding to binary 1 and 0, respectively.

Here, E_b is the energy per symbol also called as energy-per-bit E_s, T_b is the bit duration, T_s is the symbol duration, $N_0/2$ is the noise power density (W/Hz), P_b is the probability of bit-error, and P_s is the probability of symbol error.

If the sinusoidal carrier has amplitude A_c and energy per bit $E_b = 1/2 \, (A_c)2 \, T_b$, then the transmitted BPSK signal is given by

$$S_{BPSK} = \sqrt{\frac{2E_b}{T_b}} \cos\left(2\pi \ fc \ \tau + \pi + \theta c\right) \tag{5.1}$$

$$S_{BPSK} = \sqrt{\frac{2E_b}{T_b}} \cos\left(2\pi \ fc \ \tau + \theta c\right) \ 0 \le t \le T_b \left(binary \, 0\right) \tag{5.2}$$

This modulation is the most robust of all the PSKs since it takes the highest level of noise or distortion to make the demodulator reach an incorrect decision. It is, however, only able to modulate at 1 bit/symbol and so is unsuitable for high data-rate applications. BPSK is functionally equivalent to 2-QAM modulation. The BPSK signal is equivalent to a double sideband suppressed carrier amplitude modulated waveform [3]. Hence, a BPSK signal can be generated using a balanced modulator. Demodulation in BPSK receiver requires reference of transmitter signal in order to properly determine phase; hence, it is necessary to transmit carrier along with signal. It requires complex and costly receiver circuitry. It gives good BER for low SNR giving power efficiency.

5.2.2 Quadrature Phase Shift Keying

Quadrature phase shift keying (QPSK) has twice the bandwidth efficiency of BPSK. For every single modulation symbol, two bits are transmitted. The phase of carrier takes on four equally spaced values like 0, $\pi/2$, π, and $3\pi/2$. The two modulated signals, each of which can be considered to be a BPSK signal, are summed to produce a QPSK signal. QPSK transmitters and receivers are more complicated than the ones for BPSK. However, with modern electronics technology, the penalty in cost is very moderate. As with BPSK, there are phase ambiguity problems at the receiving end, and differentially encoded QPSK is often used in practice. The QPSK signal is given by

$$SQPSK(t) = \sqrt{2E_{s/T_s}} \cos\left(2\pi f_c t + (i-1)\pi/2\right) \quad 0 \le t \le T_s \ \text{for} \ i = 1,2,3,4 \tag{5.3}$$

where T_s is symbol duration and is equal to twice the bit period.

If basis function $\varphi_1(t) = \sqrt{2/Ts} \, \cos(2\pi f_c t)$

$$\varphi_2(t) = \sqrt{2/Ts} \, \sin(2\pi f_c t) \tag{5.4}$$

are defined for $0 \le t \le T_s$.

For QPSK signal set

$$S_{QPSK}(t) = \sqrt{E_s} \cos\left[\frac{(i-1)\pi}{2}\right] \varphi 1(t) - \sqrt{E_s} \sin\left[\frac{(i-1)\pi}{2}\right] \varphi 2(t) \ \text{for} \ i = 1,2,3,4 \tag{5.5}$$

QPSK has two-dimensional constellation diagram with four points. The distance between adjacent points in constellation is $\sqrt{2E_s}$. Each symbol consists of two bits; hence, $E_s = 2E_b$, then the distance between two neighboring points in QPSK constellation is given by $2\sqrt{E_b}$.

5.2.3 Offset QPSK

Offset quadrature phase-shift keying (OQPSK) is a variant of phase-shift keying modulation using four different values of the phase to transmit. It is sometimes called staggered quadrature phase-shift keying (SQPSK). Taking four values of the phase (two bits) at a time to construct a QPSK symbol can allow the phase of the signal to jump by maximum of 180° at a time. When the signal is low-pass filtered (as is typical in a transmitter), these phase shifts result in a large amplitude fluctuation. This is an undesirable quality in communication systems. By offsetting the timing of the odd and even bits by one bit-period, or half a symbol-period, the in-phase and quadrature components will never change at the same time. From the constellation of OQPSK, it can be seen that this will limit the phase shift to no more than 90° at a time. This results in much lower amplitude fluctuations than non-offset QPSK and is sometimes preferred in practice. OQPSK ensures that there are less baseband signal transitions applied to the radio frequency (RF) amplifier, which helps to remove spectrum regrowth after amplification. In OQPSK, the maximum phase shift of the transmitted signal at any time instant is limited to ±90°. Hence, hard limiting or nonlinear amplification OQPSK signal does not regenerate the high frequency side lobes as that in QPSK. It results in reduced spectral occupancy and allows more efficient RF amplification. OQPSK is very attractive for mobile communication systems where bandwidth efficiency and efficient nonlinear amplifiers are critical for low power drain.

5.2.4 M-Ary QAM

QAM is a signal in which two carriers shifted in phase by 90° are modulated and the resultant output consists of both amplitude and phase variations. As both amplitude and phase variations are present, it is also considered as a mixture of amplitude and phase modulation. Digital formats of QAM are often referred to as "Quantized QAM," and they are being increasingly used for data communications in radio communications systems. When using QAM, the constellation points are normally arranged in a square grid with equal vertical and horizontal spacings.

The most common forms of QAM use a constellation with the number of points equal to a power of 2, i.e., 2, 4, 8, 16. By using higher-order modulation formats, it is possible to transmit more bits per symbol. As the points are closer together, they are therefore more susceptible to noise and data errors. When the states are closer together, a lower level of noise can move the signal to a different decision point. QAM contains an amplitude component, hence linearity is necessary. The linear amplifiers are less efficient and consume more power, and this makes them less attractive for mobile applications. QAM is widely used in many digital data radio communications and data communications applications. Some of the more popular forms are 16 QAM, 32 QAM, 64 QAM, 128 QAM, and 256 QAM. Here the figures indicate the number of points on the constellation. QAM is a higher-order form of modulation, and, therefore, it is able to carry more bits of information per symbol. A higher-order format of QAM gives the higher data rate for the link. The general form of an M-Ary QAM signal can be defined as

$$S_i(t) = \sqrt{\frac{2E_{\min}}{T_s}} \, a_i \, \cos(2\pi \, f_c \, t) + \sqrt{\frac{2E_{\min}}{T_s}} \, b_i \, \sin(2\pi \, f_c \, t) \quad 0 \leq t \leq T \quad i = 1, 2, \ldots, M$$

(5.6)

where $E_{\min}$ is the energy of signal with lowest amplitude and a_i and b_i are pairs of independent integers chosen as per the location of particular signal point.

5.3 Orthogonal Frequency Division Multiplex

OFDM is a special case of multicarrier transmission, where a single data stream is transmitted over a number of lower-rate subcarriers (SCs). It consists of number of closely spaced modulated carriers. The modulation produces a number of overlapping sidebands spreading out on either side. The sidebands can be received without interference, since they are orthogonal to each other. The carrier spacing is made equal to symbol period. The lower data rate reduces the interference from reflections. This is achieved with the help of guard interval. OFDM increases robustness against frequency selective fading or narrowband interference.

The OFDM transmission scheme has the following key advantages:

- OFDM is an efficient way to deal with multipath; for a given delay spread, the implementation complexity is significantly lower than that of a single-carrier system with an equalizer.
- In relatively slow time-varying channels, it is possible to enhance capacity significantly by adapting the data rate per subcarrier according to the signal-to-noise ratio (SNR) of that particular SC.
- It is robust against narrowband interference because such interference affects only a small percentage of the SCs. OFDM makes

TABLE 5.2

Applications of Different Modulation Schemes

Modulation Format	Application
MSK, GMSK	GSM, CDPD
BPSK	Deep space telemetry, cable modems
QPSK, μ/4 DQPSK	Satellite, CDMA, NADC, TETRA, PHS, PDC, LMDS, DVB-S, cable (return path), cable modems, TFTS
OQPSK	CDMA, satellite
FSK	GFSK DECT, paging, RAM mobile data, AMPS, CT2, ERMES, land mobile, public safety
8, 16 VSB	North American digital TV (ATV), broadcast, cable
8PSK	Satellite, aircraft, telemetry pilots for monitoring broadband video systems
16 QAM	Microwave digital radio, modems, DVB-C, DVB-T
32 QAM	Terrestrial microwave, DVB-T
64 QAM	DVB-C, modems, broadband set top boxes, MMDS
256 QAM	Modems, DVB-C (Europe), Digital Video (US)

single-frequency networks possible, useful for broadcasting applications. It is more sensitive to frequency offset and phase noise.

- OFDM has a relatively large peak-to-average-power ratio, which tends to reduce the power efficiency of the RF amplifier.

The different applications of the modulation techniques discussed till now are enlisted in Table 5.2.

5.4 BER in Presence of Additive White Gaussian Noise

The wireless standard IEEE 802.16, WiMAX, is used for analysis which uses modulation techniques like QPSK, QAM-16, and QAM-64 on OFDMA carrier support. The transmitter and receiver are considered ideal and additive white Gaussian noise (AWGN) is introduced by channel. The maximum transmitted power is constant, regardless of the type of modulation used. It shows that a higher SNR is required to demodulate the signal within the same BER as the modulation number of bits per symbol increases. In Figure 5.1, the BER versus SNR for different modulation techniques are presented.

Table 5.3 shows the maximum BERs that can be accepted for usual services.

Muhamad Islam et al. have modeled the transceiver in MATLAB, and BPSK transmitter is used along with AWGN channel and BPSK receiver. The PSK modulation scheme for SDR is proposed to pick the constellation size that offers the best reconstructed signal quality for each average SNR.

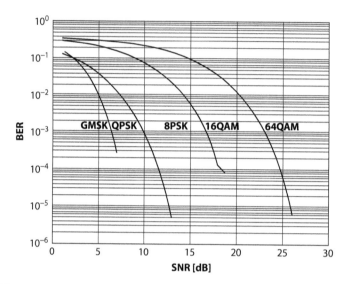

FIGURE 5.1
BER vs. SNR for different modulation techniques. (From R. Bhambare and R.D. Raut, *Int. J. Comput. Netw. Wireless Commun. (IJCNWC)*, 3, 292–299, 2013.)

TABLE 5.3

Service	Max BER
Audio	$10e^{-4}$
Video	$10e^{-5}$
Data	$10e^{-6}$
Mobile Video	$10e^{-8}$

The audio signal transmission quality is evaluated and the performance of the linear modulation is compared. It shows for a given SNR, simpler modulation schemes tend to have higher quality, giving lower bit rates. BPSK has better quality for a given SNR as compared to other modulation schemes. Therefore, it is used as the basic mode for each physical layer. It has the maximum coverage range among all transmission modes.

5.5 Adaptive Modulation and Coding

Sami H.O. Salih et al. have implemented adaptive modulation and coding technique using MATLAB [4]. The different order modulations are combined with different coding schemes. It gives higher throughput and better spectral efficiency by sending more bits per symbol. Here the various modulation

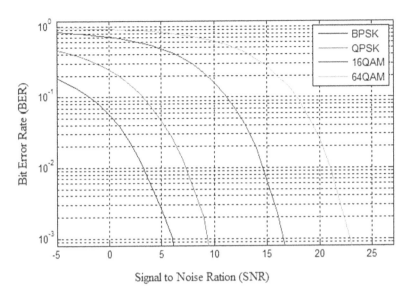

FIGURE 5.2
SNR vs. BER for different modulation schemes. (Sami H.O. Salih et al.)

types are implemented using single MATLAB function that can be called with the appropriate coefficients. Figure 5.2 shows the simulation plot of BER vs. SNR for different modulation techniques for broadband wireless access system using WiMAX.

In traditional communication systems, the transmission is designed for "worst-case" channel scenario, giving an error rate below fixed limit. The adaptive transmission has advantage of changing transmitted power level, symbol rate, coding scheme, constellation size, or any combination of these parameters in order to deliver better link average spectral efficiency given by bits/sec/Hz.

5.6 M-QAM for Digital Radio and Television Broadcasting (DVB-T, DVB-T2)

Mohamed Al Wohaishi et al. have shown the analysis of digitally modulated signals in communication systems, which use software-defined radio concept and modern synthetic instruments [5]. M-QAM is used for transmission of information in DVB-T, DVB-T2.

Simple picture is transmitted through simulated radio channel to show the result of signal impairments. Modular PXI HW platform was used in connection with graphically oriented development environment [6]. Terrestrial DVB-T broadcasting uses QPSK, 16-QAM, and 64-QAM modulation schemes, while terrestrial DVB-T2 broadcasting, which allows transmission of high definition picture format, uses 256-QAM modulation scheme. The simulation of different M-QAM modulation shows that increase of the state number leads to an increase of transfer rate (transfer more bits per symbol). The downside, however, is that with the growing number of states, BER increases at the same transmission power as a result of worse distribution of symbols in constellation diagram (Figure 5.3).

The commonly used QAM techniques are 16 QAM and 64 QAM. It represents the best trade-off between theoretical performance and implementation complexity. Michel Borgne made a comparative study of four 2^n state QAM techniques. The effects of filtering, interference, amplifier nonlinearities, and selective fading are analyzed. The spectral efficiency of QAM is attractive, but it is difficult to implement. These schemes are sensitive to nonlinearities and selective fading. In such systems, it is necessary to include compensative devices such as nonlinearity cancellers and adaptive equalizers. QAM is bandwidth efficient but requires strong signal strength for good BER. This is particularly true for the denser bandwidth schemes such as 64 QAM.

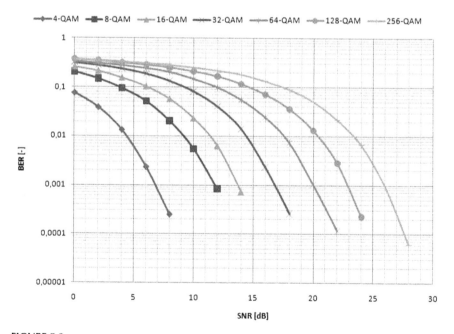

FIGURE 5.3
Measured BER dependency on SNR. (From Wohaishi, M.A. et al., Analysis of M state digitally modulated signals in communication systems based on SDR concept, *Proceedings of the 6th IEEE International Conference on Intelligent Data Acquisition and Advanced Computing Systems*, Prague, 2011, pp. 171–175.)

5.7 Configurable Architecture of Modulation Technique for SDR

Jignesh Oza et al. have elaborated the configurable architecture of modulation technique for SDR application. The configurable architecture eliminates the need for hardware changes when the technology is upgraded. Modulation techniques are implemented on FPGA and MATLAB in order to optimize the SDR architecture. It is shown that modulation technique FSK and QPSK when configured together gives maximum optimization with good performance. This chosen combination has maximum common hardware as well, along with quite few other features over other combinations like ASK+MSK, MSK+FSK, and QAM+FSK. Table 5.4 elaborates the different features shown by QPSK+FSK combination.

SDR systems are required to support multiple air interfaces and signal processing functions at the same time. Byeong-Gwon Kang discussed the modulation schemes of OCQPSK (orthogonal complex QPSK) and HPSK (hybrid PSK) used for IMT-2000 Services of synchronous systems and asynchronous systems, respectively. With growing multimedia applications including data services in cellular networks, wireless Internet access, and wireless LANs, system flexibility is required for high-speed mobile radio system. As different modulation schemes and frequency bands are required for different services, SDR can replace hardware device by upgradable software programmable devices. Therefore, SDR can offer good choice for adaptive modulation or multiple access schemes. OCQPSK and HPSK modulation schemes are adopted in IMT-2000 Services of CDMA 2000 and WCDMA, respectively. Each modulation method can be chosen by external selection in the implemented modem with SDR applications.

TABLE 5.4

Configurable Modulation Techniques Features

Features	Modulation Technique (QPSK+FSK)
Bandwidth	Only highest bandwidth is that of the FSK signal which is very less compared to other techniques used.
SNR	Higher SNR
Data rate	Higher data rate
Applications	Can be used in wide applications
PB vs. Eb/No	As QPSK is combined, requires lesser Eb/No for a given value of PB
Reliability	Higher
Hardware complexity	Less complex
Hardware cost	Costly
Implementation on FPGA	Less complex compared to MSK+FSK and QAM+FSK

Source: Oza, J. et al., Optimized configurable architecture of modulation techniques for SDR applications, *International Conference on Computer and Communication Engineering (ICCCE'10)*, Kuala Lumpur, IEEE, 2010, pp. 1–5.

5.8 GMSK Modulation

GMSK has been widely used in mobile wireless communication due to its constant envelope signal feature, which ceases the requirement for power amplifier linearity. In this case, the phase of the carrier is instantaneously varied by "Modulating" signal. It is used as modulation standard of GSM system. It can be regarded as two-level FSK modulation with modulation index of 0.5. Jagadeesh Gurugubelli et al. have used the linear approximated GMSK in SDR environment because it gives a common I/Q modulator that can be used for all second-generation systems. The generalized parametrizable modulator for a reconfigurable radio can perform GMSK and QPSK modulations. GMSK is the underlying modulation scheme for global system for mobile (GSM) standard, while QPSK technique is the basic scheme for code division multiple access (CDMA) standard. Harada and Masayki presented SDR that can realize global positioning service (GPS) navigation system, vehicle information and communication system (VICS), electronic toll collection system (ETC), AM/FM radio broadcasting services, and FM multiplex broadcasting system. It also served modulation schemes such as BPSK, QPSK, GMSK, ASK, and $\pi/4$ QPSK. The SDR realizes simultaneous multiple services when user would like to use several communication services in driving situations.

5.9 Orthogonal Frequency Division Multiplexing

The next-generation wireless communication systems require higher data rates transmission for better quality of service. Multiple antennas and orthogonal frequency division multiplexing (OFDM) are mostly favored technologies for 3G and 4G. IEEE and ETSI have selected OFDM as their physical layer techniques for next generation of wireless systems.

High data rate communication systems are restricted by problem of intersymbol interference (ISI) due to multiple paths. OFDM is considered as most promising technique to combat this problem. OFDM is very efficient in spectrum usage and is very effective in a frequency selective channel. OFDM is already used in digital audio and video broadcasting and wireless LAN's (802.11 Family) and is emerging as a technology for future broadband access. Haitham J. Taha et al. have presented combination of OFDM and CDMA techniques. It offers great advantage, which can lower the symbol rate in each subcarrier. The longer symbol duration makes it easier to synchronize the transmission. The main advantages of multicarrier modulation are that it solves multipath propagation problem using simple equalization at the receiver. This system is more efficient than single carrier transmission and supports multiple access systems such as (TDMA, FDMA, MC-CDMA) and various modulation techniques. Weinstein and Elbert proposed a complete OFDM system which includes generation of signals with an FFT and adding generated intervals in case of multipath channels. M.A. Lard and Lassalle discussed OFDM for broadcast application and mobile reception.

5.10 Comparison of Representative Modulation Schemes

In this section, the modulation schemes that are used for SDR communication are compared with respect to their performance under a variety of digital radio channels conditions. A performance measure used is the baseband equivalent E_b/N_o (defined as the ratio of average signal energy per bit to noise power spectral density, as measured at the input to the receiver) required to achieve a BER of 10e^{-4}. This error rate is adequate for most general-purpose digital radio applications.

5.10.1 Ideal Performance

In order to establish a baseline for comparison, Table 5.5 presents the ideal performance of the representative modulation techniques in the presence of AWGN. The performances shown by QAM, MSK, and QPSK are almost identical. The MSK and OQPSK differ only in the weighting functions applied to I and Q channels.

TABLE 5.5

Ideal Performance of Representative Modulation Schemes

Modulation Scheme	E_b/N_o Required[a]
QAM	8.4
FSK Non-coherent detection $d = 1$	12.5
MSK $d = 5$	8.4
MSK $d = 5$ differential encoding	9.4
BPSK—Coherent detection	8.4
DPSK	9.3
QPSK	8.4
DQPSK	10.7
OQPSK	8.4
8-ary PSK—Coherent detection	11.8
16-ary PSK—Coherent detection	16.2

[a] For BER of 10e^{-4}.

5.10.2 Spectral Characteristics

The spectral characteristics of the modulation schemes can be compared according to the extent to which a signal will interfere with signals of adjacent channels. This can be measured by the attenuation of a signal's power spectrum a specified distance from the center frequency. In case of phase modulation systems, if phase transitions can be made to occur more smoothly, improved spectral characteristics can be achieved. The side lobes can always be reduced by suitable, postmodulation filtering, which results in the degradation in performance. Thus, the spectral merits of the various schemes can only be judged after doing a detailed study of the trade-offs between cost and performance. Another important spectral property is the bandwidth required to transmit at a specified information rate. The so-called "speed" of a modulation technique (equal to R/W, where R is the data rate and W is the IF bandwidth) is an important figure of merit. In Table 5.6, the speeds for each technique are listed together with the E_b/N_o required for a BER when the signal is filtered at the indicated bandwidth (i.e., the degrading effects of finite bandwidth are included).

5.10.3 Effects of Interference

The effect of co-channel and adjacent channel interference is important factor in evaluating potential modulation schemes for digital radio. It can be observed that MSK scheme has large advantage over the AM and PM schemes when no postmodulation filtering is employed. Noncoherent FSK and BPSK show the minimum degradation from ideal performance, while the 8-ary and 16-ary schemes show the maximum degradation.

TABLE 5.6

Relative Signaling Speeds of Representative Modulation Schemes

Modulation Scheme	Speed (bits/s per Hz)	E_b/N_o Required[a]
QAM	1.7	9.5
FSK Non-coherent detection $d = 1$	0.8	11.8
MSK $d = 5$	1.9	9.4
MSK $d = 5$ Differential encoding	1.9	10.4
BPSK—Coherent detection	0.8	9.4
DPSK	0.8	10.6
QPSK	1.9	9.9
DQPSK	1.8	11.8
8-ary PSK—Coherent detection	2.6	12.8
16-ary PSK—Coherent detection	2.9	17.2

[a] For BER of 10e−4 and d = FM Modulation Index.

TABLE 5.7

Performance of Representative Modulation
Schemes in the Presence of CW Interference

Modulation Scheme	E_b/N_o Required[a]	
FSK Non-coherent detection $d = 1$	14.7	13.3
BPSK—Coherent detection	10.5	9.2
DPSK	12.0	10.3
OPSK	12.2	9.8
DQPSK	>20	14.0
8-ary PSK—Coherent detection	~20	15.8
16-ary PSK—Coherent detection		>24

[a] For BER of 10e^{-4} and d = FM Modulation Index.

5.10.4 Effects of Fading

Fading is another problem often encountered on digital radio links. If the
fading is caused by two resolvable multipath components, then the results of
Table 5.7 can be utilized (the CW interferer can represent the signal from the
secondary path). If the fading is caused by a large number of equal amplitude components, the Rayleigh fading model is more useful.

5.10.5 Effects of Delay Distortion

Most of the delay distortion observed on line-of-sight radio links is introduced by the radios and not the channel. For quadratic and linear delay
distortion for the case in which the maximum differential delays (relative
to the mid-band delay) is equal to the symbol duration. DQPSK suffers
severe degradation from quadratic delay distortion, while the coherent bi-orthogonal schemes (QAM, MSK, and the variations of QPSK) are
degraded significantly by linear delay distortion. Thus, delay distortion
can be an important criterion in the selection of a modulation scheme for
digital radio.

5.11 SDR-Based Modulator Design and Implementation Using GNU Radio

Software-defined radios are alternative form of radio systems [7]. In
hardware-defined radios, typical components such as mixers, filters, modulators, and so on are implemented in hardware; thus, hardware-defined
radios cannot be reconfigured in runtime by the user to switch to different

vacant channels as per requirement. To satisfy this need of cognitive system, software-defined radios are used where we can configure the components in runtime as per the need and available channel. In software-defined radios, the components are implemented on FPGA boards, ASIC boards, or other general-purpose boards as per the application. The fundamental characteristics of SDR are that the software defines the transmitted waveform and the demodulated received signal. This offers a great flexibility for wireless communication. A variety of tools, algorithms, and protocols can be implemented and verified easily in the same manner as if they were done in a test bed constructed by real radio platforms.

The implementation of SDR takes large amount of time, effort, and cost, as it has to process tremendous amount of data in real time. MATLAB and other similar tools cannot process such huge amount of data in real time and inculcate large expense in acquiring licenses. Open source software has been developed to solve this problem. The most widely used and accepted software is GNU Radio [8,9]. It offers various building blocks for signal processing. It offers methods to manipulate data flow between the blocks. Moreover, it provides protection to the system from damage due to high-speed reading and writing operations and implementing high sample rate devices via "Throttle" block. The advantages of using GNU Radio are as follows:

1. Inbuilt blocks, which are directly used for the designing of a system
2. Provision for adding a self-constructed block
3. Any system as a whole can be implemented using the software

GNU Radio offers various functions such as mathematical operations, logical operations, FFT/IFFT blocks, filters, and so on. It also offers type conversion such as float to short block, integer to float block, and complex to real block. Different sources and sinks are also offered, which thus makes it easy to build any system. GNU component blocks are designed using C++ and connected using python.

An example of such abstracted flow graph is given in Figure 5.4, which shows data flow between various blocks.

The most common hardware used with the GNU Radio to build an SDR system is Universal Software Radio Peripheral (USRP) [10,11]. It consists of two main subdevices, a mother board and daughter boards, which can convey and/or receive data at different frequencies. The daughter boards can be easily exchanged, which provides more flexibility to the system. The mother board consists of FPGA and their main function is to convert analog signals into baseband digital signal and vice versa, thus needing ADC and DAC. To solve the issue of data realization by ADCs and DACs at very high speed, the daughter boards are introduced in the USRP.

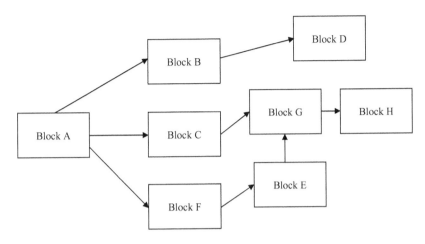

FIGURE 5.4
Data flow graph for GNU Radio. (From DucToan Nguyen, *Thesis of Implementation of OFDM using GNU and USRP*, University of Wollongong, 2013.)

5.11.1 Design of Amplitude Modulator

In amplitude modulation, the amplitude (signal strength) of the carrier wave is varied in proportion to the waveform being transmitted. That waveform may, for instance, correspond to the sounds to be reproduced by a loudspeaker, or the light intensity of television pixels. Implementation of amplitude modulation is shown in Figure 5.5. The data flow graph has been designed by considering the following equations. The carrier wave of frequency f_c and amplitude A is given by:

$$c(t) = A \times \sin\left(2\pi f_c t\right)$$

where $f_c = 10$ kHz and $A = 1$.
   The modulation waveform of frequency f_m and amplitude M is given by:

$$m(t) = M \times \cos\left(2\pi f_m t\right)$$

where $f_m = 1$ kHz and $M = 0.7$.
   The amplitude-modulated wave is obtained by considering the following equation:

$$y(t) = \left[1 + m(t)\right] \times c(t)$$

The output waveform obtained for amplitude modulator using GNU Radio for the given specification is shown in Figure 5.6.

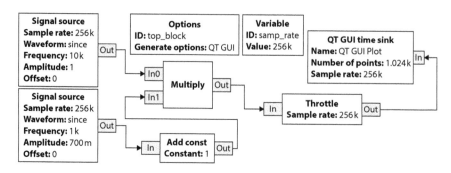

FIGURE 5.5
Data flow representation of amplitude modulator. (From O. Chatterjee and R. Raut, *Inter. Conf. Recent Trends Engine. Sci.*, 2017.)

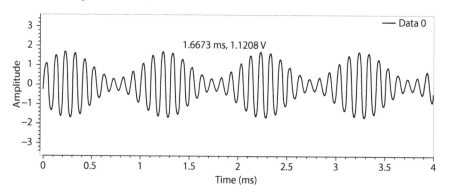

FIGURE 5.6
Output waveform of amplitude modulator using GNU Radio. (From O. Chatterjee and R. Raut, *Inter. Conf. Recent Trends Engine. Sci.*, 2017.)

5.11.2 Design of SSB Modulator

SSB modulation is an improvement over amplitude modulation, which uses transmitter power and bandwidth more efficiently. Amplitude modulation produces an output signal that has twice the bandwidth of the original baseband signal. SSB modulation avoids this bandwidth doubling, and the power is wasted on a carrier, at the cost of increased device complexity and more difficult tuning at the receiver.

The data flow graphs of the SSB modulator are designed using the following equations

$$S_{LSB}(t) = s(t) \times \cos(2\pi f_c t) + \hat{s}(t) \times \sin(2\pi f_c t)... \text{ For lower sideband}$$

$$S_{USB}(t) = s(t) \times \cos(2\pi f_c t) - \hat{s}(t) \times \sin(2\pi f_c t)... \text{ For upper sideband}$$

where $s(t)$ is the information signal, $\hat{s}(t)$ is the Hilbert transform of the data signal, and f_c is the carrier frequency

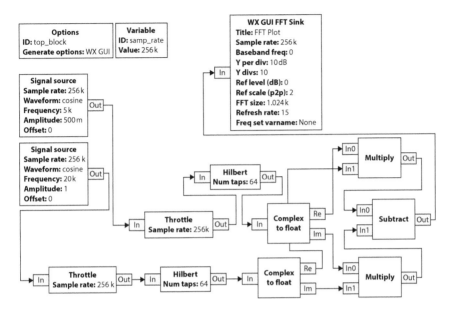

FIGURE 5.7

Data flow representation of upper sideband modulator. (From O. Chatterjee and R. Raut, *Inter. Conf. Recent Trends Engine. Sci.*, 2017.)

$$s(t) = \cos(2\pi f_m t)$$

For the upper sideband modulator, the data flow graph is shown in Figure 5.7.

The output waveform obtained for upper sideband modulator using GNU Radio for the given specification is shown in Figure 5.8.

For the lower sideband modulator, the dataflow graph is shown in Figure 5.9.

The output waveform obtained for lower sideband modulator using the GNU Radio for the given specification is shown in Figure 5.10.

5.11.3 Demodulation of USRP-Generated .DAT File

A data file is a file that contains several seconds of recorded signals from the AM broadcast band. This data file is obtained from a USRP. The .DAT file used contains a set of two recorded data of audio files at two different frequency bands. The amplitude-modulated signal from the .DAT file is demodulated using amplitude demodulation technique. Two sliders are used in the design: one to adjust the volume of the audio files and the other to slide to different frequencies to listen to both the audio files stored at different frequencies.

The design flow graph is shown in Figure 5.11.

The output waveform obtained for demodulation of USRP-generated .DAT file using GNU Radio for the audio file at 710 kHz is shown in Figure 5.12.

The output waveform obtained for demodulation of USRP-generated .DAT file using GNU Radio for the audio file at 790 kHz is shown in Figure 5.13.

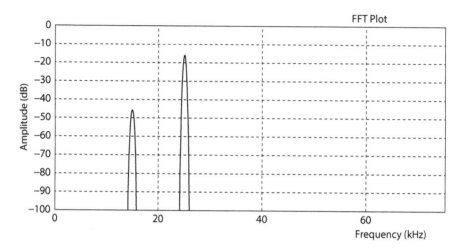

FIGURE 5.8

Output waveform of upper sideband modulator using GNU Radio in frequency domain. (From O. Chatterjee and R. Raut, *Inter. Conf. Recent Trends Engine. Sci.*, 2017.)

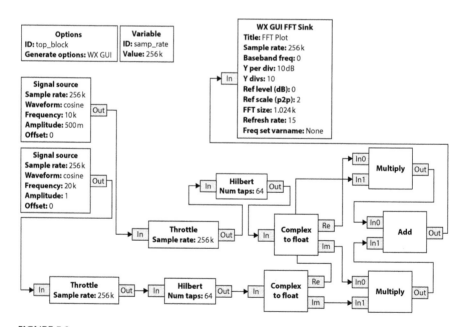

FIGURE 5.9

Data flow graph representation of lower sideband modulator. (From O. Chatterjee and R. Raut, *Inter. Conf. Recent Trends Engine. Sci.*, 2017.)

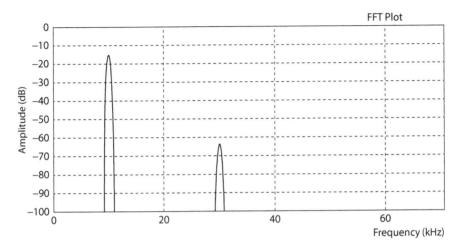

FIGURE 5.10
Output waveform of for lower sideband modulator using GNU Radio in frequency domain.
(From O. Chatterjee and R. Raut, *Inter. Conf. Recent Trends Engine. Sci.*, 2017.)

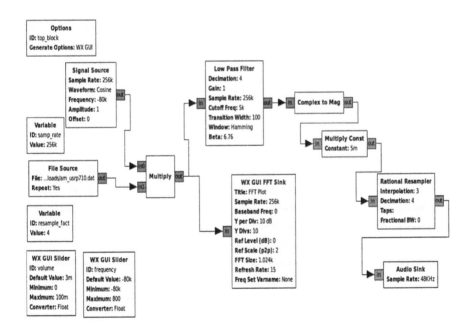

FIGURE 5.11
Data flow graph representation of demodulation of USRP-generated .DAT File. (From
O. Chatterjee and R. Raut, *Inter. Conf. Recent Trends Engine. Sci.*, 2017.)

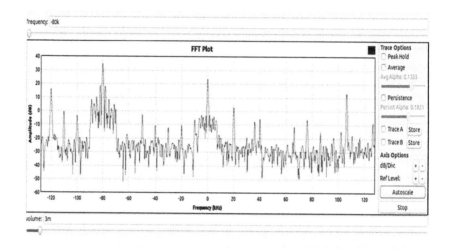

FIGURE 5.12
Demodulation of USRP-generated .DAT file using GNU Radio for the audio. (From O. Chatterjee and R. Raut, *Inter. Conf. Recent Trends Engine. Sci.*, 2017.)

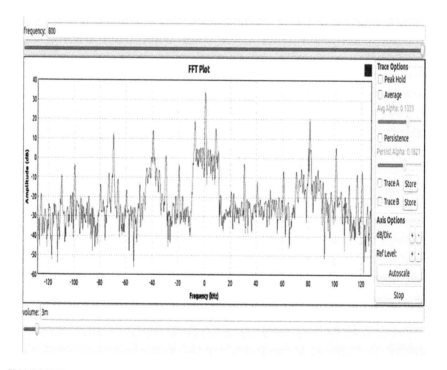

FIGURE 5.13
Demodulation of USRP-generated .DAT file using GNU Radio for the audio file at 790 kHz. (From O. Chatterjee and R. Raut, *Inter. Conf. Recent Trends Engine. Sci.*, 2017.)

5.12 Conclusions

This chapter presents an analysis of the modem modulation techniques that are used in the latest wireless standards, such as IEEE–802.11 and IEEE–802.16 also known as WiMAX. It also gives an insight into selecting proper modulation techniques for SDR as per existing channel quality. SDR system is a useful and adaptable future proof solution to cover both existing and emerging standards. It provides the designs with reconfigurability, intelligence, and software programmable hardware. SDR has given promising solution in building multimode, multiband, and multifunctional wireless communication devices. The quality of service provided by wireless communication services can be greatly improved with the help of correct selection of modulation technique. It will serve to increase radio coverage and reduce power consumption. The better return on investment will be provided for next-generation wireless communication systems along with good quality. Further, in this chapter, amplitude modulator, SSB modulator, and demodulation of USRP-generated .DAT file are implemented with the help of predefined general signal processing blocks present in GNU Radio software. It has been observed that GNU Radio provides high flexibility and ease in designing signal processing blocks with its main feature that allows processing real-time data with high sampling rate and fast computation over other signal processing software. All of the above modulator and demodulation designs will help in understanding and improving the performance parameters of SDR.

References

1. R. Bhambare, R. Raut, 2013. A survey on digital modulation techniques for software defined radio applications. *International Journal of Computer Networks and Wireless Communications (IJCNWC)*, 3(3): 292–299.
2. R. Pandey, K. Pandey, 2014. An introduction of analog and digital modulation techniques in communication. *Journal of Innovative trends in Science, Pharmacy & Technology*, 1(1): 80–85.
3. G. Tharakanatha, S.K. Mahaboob, V.B. Chanda, I. Hemalatha, 2013. Implementation and bit error rate analysis of BPSK modulation and demodulation technique using MATLAB. *International Journal of Engineering Trends and Technology (IJETT)*, 4(9): 4010–4014.
4. S.H.O. Salih, M.M.A. Suliman, 2011. Implementation of adaptive modulation and coding techniques using Matlab. *Proceedings ELMAR-2011*, Zadar, pp. 137–139.
5. M.A. Wohaishi, J. Zidek, R. Martinek, Analysis of M state digitally modulated signals in communication systems based on SDR concept. *Proceedings of the 6th IEEE International Conference on Intelligent Data Acquisition and Advanced Computing Systems*, Prague, 2011, pp. 171–175.

6. R. Martinek, M. Al-Wohaishi, J. čídek, Software based flexible measuring systems for analysis of digitally modulated systems. *9th RoEduNet IEEE International Conference*, Sibiu, 2010, pp. 397–402.

7. O. Chatterjee, R. Raut, SDR based modulator design and implementation using GNU radio, *International Conference on Recent Trends in Engineering and Science* (ICRTES 2017), 20–21 January 2017.

8. Guided tutorial on GNUradio. Available: http://gnuradio.org/redmine/projects/gnuradio/wiki/TutorialsCoreConcepts.

9. GNU Radio Companian, Available: http://gnuradio.org/redmine/projects/gnuradio/wiki/GNURadioCompanion.

10. D.T. Nguyen, 2013. *Thesis of Implementation of OFDM using GNU and USRP.* University of Wollongong.

11. D.C. Tucker, G.A. Tagliarini, 2009. *Tagliarini, Prototyping with GNU Radio and the USRP—Where to Begin, Proceeding.* IEEE Southeastcon, Atlanta, GA, pp. 50–54.

6

Spectrum Sensing in Cognitive Radio Networks

6.1 Introduction

Spectrum sensing in general refers to the detection of the occurrence of the primary or incumbent users in an accredited spectrum and is an essential problem for cognitive radio. Spectrum sensing has therefore reborn as an exceptionally active research area in recent years in spite of its long history.

The fundamental idea of cognitive radio is to reuse the spectrum or share it in an opportunistic manner, which allows the secondary users to converse over the spectrum allocated/accredited to the incumbent users when they are not using it [1]. To accomplish this task, it is mandatory for secondary users to recurrently carry out spectrum sensing, in other words, detect the presence of the primary users. Whenever the primary users become active and start to communicate on the licensed band, the secondary users have to become aware of the presence of the incumbent user and relinquish the channel or reduce transmit power within certain amount of time. Moreover, the secondary user must assure that it does not cause any intrusion to the primary user. In IEEE 802.22 standard, it is necessary for the secondary users to become aware of the TV and wireless microphone signals (primary users) and evacuate the channel within two seconds once they sense the presence of the primary user [2].

Furthermore, for TV signal detection, it is essential to attain 10% probability of false alarm and 90% probability of detection at signal-to-noise ratio (SNR) level as low as −20 dB. There are number of factors that make spectrum sensing practically exigent. The required SNR for detection may be very low. For example, consider that primary transmitter is in the close proximity of a secondary user (the detection node) and the broadcasted signal of the primary user can be deep faded, and, as a result, the primary signal's SNR at the secondary receiver is less than −20 dB. Nonetheless, the secondary user still needs to sense the presence of the primary user and stay away from using the channel because it may strongly get in the way of the primary receiver if it broadcasts.

A realistic scenario of this is a wireless microphone operating in TV bands, which only broadcasts with a power less than 50 mW and a bandwidth less than 200 KHz. If a secondary user is several hundred meters away from the microphone device, the received SNR may be well below −20 dB. Moreover, multipath fading and time dispersion of the wireless channels make difficult the sensing problem. Multipath fading may cause the signal power to rise and fall as much as 30 dB. On the other hand, unknown time dispersion in wireless channels may turn the coherent detection erratic. Third, the noise/intrusion level may alter with time and location, which yields the noise power ambiguity issue for detection. Facing these challenges, spectrum sensing has attracted a lot of researchers to pursue research over recent years, although it has a long history.

Broadly speaking, spectrum sensing is a method of acquiring the information concerning through spectrum usage characteristics across manifold dimensions such as space, frequency, time, and code. It also comprises of the compilation of information regarding the type of signals, bandwidth, their modulation, carrier frequency, and so on occupying the spectrum.

6.2 Conventional Methods of Spectrum Sensing

In literature, three main signal processing techniques are used to capture the signals of interest. They are as follows [3]:

1. Energy detection
2. Cyclostationary feature detection
3. Matched filter detection

6.2.1 Energy Detection

Energy detector is simple in structure and most common way of spectrum sensing technique because of its low computational time and low implementation complexities [4]. Energy detector does not need any knowledge regarding primary user's signal structure. In this method, signals are detected by comparing the energy detector output with a threshold value. In energy detector, major problem is the selection of threshold value, which is essential for detecting primary channel user. Probability of detection (P_d) and probability of false alarm (P_f) are the main factors for providing the appropriate information of the availability of the spectrum.

The block diagram of energy detector is shown in Figure 6.1. The input is given to the band pass filter, with the center frequency f_s and bandwidth W.

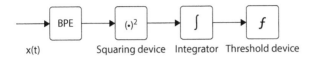

x(t)   Squaring device   Integrator   Threshold device

FIGURE 6.1
Block diagram of energy detection. (From Nimbalkar, N.O., et al., *Inter. J. Recent Innov. Trends Comp. Comm.*, 3, 703–706.)

A squaring device is used to measure the received energy, which is followed by the band pass filter after that integrator is used to determine the observation interval. Finally, the output of the integrator is compared with a threshold energy value to decide the presence of primary user.

In a nonfading environment, where h is the amplitude of the signal channel, probability of detection P_d and probability of false alarm P_f are given by following formulas

$$P_d = P(Y > \lambda / H_1) = Q_m\left(\sqrt{2\gamma}, \sqrt{\lambda}\right) \tag{6.1}$$

$$P_f = P(Y > \lambda / H_0) = \Gamma\left(m, \lambda/2\right) / \Gamma\left(m\right) \tag{6.2}$$

where Y is the SNR, $m = TW$ is the (observation/sensing) time bandwidth product Γ (.) and Γ (.,.) are complete and incomplete gamma functions, and Q_m ( ) is the generalized Marcum Q-function. In a fading environment, h is the amplitude gain of the channel that varies due to fading effect, which makes the SNR variable. P_f is the same for both fading and nonfading cases because it is independent of SNR. P_d is the probability of detection for SNR. In this case, average probability of detection may be derived by averaging over fading statistics:

$$P_d = \int xQ_m\left(\sqrt{2\gamma}, \sqrt{\lambda}\right) f\gamma(x)dx \tag{6.3}$$

where $f\gamma(x)$ is the probability distribution function of SNR under fading. A low value of P_d expresses an absence of primary user; it means that the CR user can access that spectrum. A high value of P_f expresses that there is no signal in the channel. It shows that in fading environment, different CRs need to cooperate for detecting the presence of primary user. In such conditions, a CR model is helpful for relating different parameters such as detection probability, number and spatial distribution of spectrum sensors, and, more importantly, propagation characteristics. Disadvantage of energy detection is that its performance is sensitive to noise. In energy detector, it is difficult to differentiate between signal power and noise power; it only gives the information related to the absence or presence of the primary user.

6.2.2 Matched Filter Detection

Matched filter is also called as optimum method for primary user detection when we transmit the known signal. In matched filter detector, it requires prior knowledge of the signal transmitted by the primary user. The matched filter proceeds from threshold detector and used to detect the presence of primary user. The matched filter is used for increasing the SNR ratio in the presence of additional white noise. This technique is possible only if number of users are very small. A matched filter detector is obtained by relating a known signal present in the channel, with noise present in environment for detecting the presence of the known signal in the channel, which is same as unknown signal with time of the signal. It performs two convolution functions, one function is to find out the level of similarity and another function is to find out the threshold value. Advantage of matched filter is that it requires short time to achieve high processing gain due to channel detection. Disadvantage of matched filter is that it requires a devoted sensing receiver for each primary user signal and requires the previous information of primary user signal, which is very difficult to obtain at the CRs (Figure 6.2).

6.2.3 Cyclostationary Feature Detection (CFD)

In cyclostationary detector, signals are in general form such as sine wave, pulse trains, hopping sequences, or cyclic prefixes, which results in regularity. Even if the data is in random form, these signals are characterized as cyclostationary; hence, their statistics, mean, and autocorrelation exhibit periodicity. This method is achieved by comparing a spectral function. The periodicity is beneficial for signal format so that it is helpful for the receiver for parameter estimation like pulse timing, etc. This regularity is useful for the detection of random signals with the noise and other modulated signals. From research, it is found that cyclostationary feature detector method is better as compared with simple energy detection and match filtering [7]. As discussed earlier, a matched filter detector requires previous knowledge about primary user, while an energy detector as a noncoherent detection does not require any prior knowledge about primary user. Although energy detector is easy implement, it is highly susceptible to band interference and changing noise levels, and it is difficult to differentiate between the signal power and noise power.

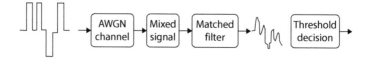

FIGURE 6.2
Block diagram of matched filter detection. (From Nimbalkar, N.O., et al., *Inter. J. Recent Innov. Trends Comp. Comm.*, 3, 703–706.)

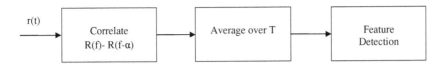

FIGURE 6.3
Block diagram of cyclostationary feature detection. (From Nimbalkar, N.O., et al., *Inter. J. Recent Innov. Trends Comp. Comm.*, 3, 703–706.)

Implementation of CFD is shown in Figure 6.3. Detected features are in the form of signals and presence of interferers. When the correlation factor is greater than the threshold, then there is a presence of primary user in radio environment. CFD performs better than energy detector and matched filter detector because it can differentiate between signal power and noise. However, it is computationally complex and therefore requires long processing time, which generally reduces the performance of cognitive radio. Signal processing techniques motivate the need to study other feature detection techniques that can improve sensing detection and type of signals in low SNR.

6.3 Cooperative Spectrum Sensing

When there are several cognitive users dispersed in diverse locations, it is promising for them to assist each other to accomplish higher sensing trustworthiness [8,9]. This will result in various cooperative sensing schemes. If each user sends its observed data or processed data to a specific user, which jointly processes the amalgamated data and makes an ultimate verdict, this cooperative sensing scheme is called *data fusion*. Alternatively, if multiple receivers process their observed data autonomously and send their assessment to a specific user, which then makes a final decision, it is called *decision fusion*. Cooperative spectrum detecting essentially builds the likelihood of detection in a cognitive radio network. Cooperative spectrum detection shown in Figure 6.4 is more precise in detection, since the issues of multipath fading and shadowing experienced by solitary optional user detection has been limited by auxiliary users offering data to each other about their individual spectrum detection.

In the cooperative (centralized) spectrum sensing, a fusion center (FC) gathers the whole spectrum detecting data from various auxiliary users and recognizes the accessible spectrum and broadcast this data to the secondary users. On the other hand, the cooperative (distributed) detection, the auxiliary users perform spectrum sensing individually and decide on their observation about the occupancy of spectrum. Moreover, they disseminate this data among themselves.

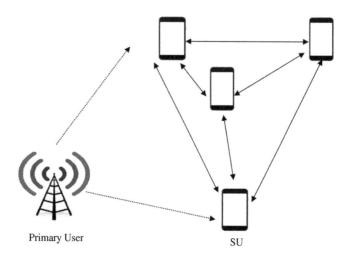

FIGURE 6.4
Cooperative spectrum sensing scheme.

Some other detection methods proposed in the literature are combined energy detection multitaper spectrum estimation, filter bank spectrum estimation, likelihood ratio test, wavelet-based sensing, covariance-based sensing, and waveform-based sensing. Waveform-based sensing or coherent sensing investigates the received signal by showing a relationship with a recognized signal pattern or a template. This approach can do better than the energy detector in convergence time and trustworthiness. Furthermore, it requires smaller sensing time, and it has been revealed that the performance of the sensing algorithm get better as the length of the signal template increases. However, this technique is vulnerable to synchronization errors and can be useful only to systems with recognized signal patterns. The other noteworthy method, wavelet-based technique can look into a wide bandwidth simultaneously. The wideband under deliberation is separated into successive frequency sub-bands wherever the power spectral characteristic inside each sub-band is smooth but display a sporadic change between adjacent sub-bands. By investigating these discontinuities, information on the location and intensities of spectrum bands can be extracted. Once the PSD edges are perceived, the powers within bands amid two edges are anticipated. Using this information and edge positions, the frequency spectrum can be characterized as busy or vacant. The type of wavelet used determines the efficiency of this system.

Even though cooperative sensing can attain enhanced performance, there are few issues associated with it. Trustworthy information exchange amid the cooperating users ought to be assured. In an ad-hoc network, this is by no means a trouble-free task. The majority of data fusion methods in literature are based on the easy energy detection and flat-fading channel model, while more sophisticated data fusion algorithms such as cyclostationary detection, eigen value-based detection, space–time combining, etc. need to be further investigated. The existing decision fusion techniques have by and large assumed that decisions of diverse users are autonomous, which may not be factual for the reason that all users in fact receive signals from various common sources. Eventually, practical fusion algorithms ought to be robust to data errors owing to channel impairment, interference, and noise.

6.4 The Two Hypothesis Model

Spectrum sensing is one of the most important tasks in the cognitive radio network. It is the first step for communication to take place, which needs to be performed. Spectrum sensing is popularly known as the hypothesis test because it is considered as identification processes. The spectrum sensing is a test of the following two hypotheses:

$$H_0 : x(t) = n(t)$$

$$H_1 : x(t) = s(t) + n(t)$$

$s(t)$ is the signal that is transmitted by the PUs.
$x(t)$ is the signal which is received by the SUs.
$n(t)$ is known as the AWGN (additive white Gaussian noise).

H_0 hypothesis is used to tell that no primary signals are present in the spectrum and only noise is present. Hence, it is allotted to the secondary users. H_1 hypothesis is used to tell that primary signals are present in the spectrum along with the noise. Hence, it is not allotted to the secondary users else there will be harmful interference to the primary users.

The fundamental objective of spectrum sensing is to use the system for decision-making process to check the availability of the spectrum holes, record the data, and then carry out simulation to analyze the stored data.

6.4.1 Probability of False Alarm (P_{fa})

Probability of false alarm occurs when no primary signals are present in the spectrum, but it appears as if the PU is present and hence bands are not allocated to the SUs. It occurs when only noise is present in the channel and energy of noise level exceeds the predefined threshold value and hence the presence of primary user is detected by the decision-making device. This is false representation and should be minimized.

$$P_{fa} = \int_{V_{TH}}^{\infty} \frac{1}{\sqrt{2\pi\sigma_n^2}} \exp\left(-\frac{v^2}{2\sigma_n^2}\right) dv \tag{6.4}$$

6.4.2 Probability of Missed Detection (P_m)

Probability of missed detection occurs when primary signals are present in the spectrum, but they are not properly identified and hence spectrum is allocated to other SUs. This causes interference to the primary users. It happens when a signal is present in the channel and energy of signal present do not exceed the predefined threshold value and hence the presence of primary users is not detected by the decision-making device. This is missed detection condition and hence should be minimized. Figure 6.5 shows the trade-off between P_m and P_{fa} with respect to the threshold value:
$P_m = 1 - P_d$

$$P_d = \int_{V_{TH}}^{\infty} \frac{1}{\sqrt{2\pi\sigma_n^2}} \exp\left(-\frac{v^2 + V_0^2}{2\sigma_n^2}\right) dv \tag{6.5}$$

In Figure 6.5, it can be observed that as the threshold value decreases, there is decrease in the probability of missed detection; however, it would increase the probability of false alarm. On the other hand, by increasing the threshold value, the probability of false alarm decreases, whereas there is an increase in probability of missed detection. Since both are unwanted and both cannot be decreased simultaneously, trade-off between these two parameters is done and the threshold is set accordingly.

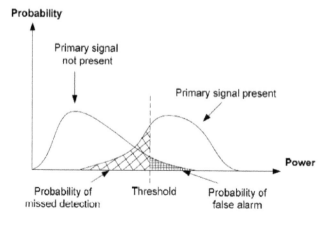

FIGURE 6.5
PDFs for hypothesis test model. (From Khobragade, A.S. and Raut, R.D., *Inter. J. Elect. Comput. Eng. (IJECE)*, 7, 2683–2695, 2017.)

6.5 Hybrid Spectrum Sensing Method for Cognitive Radio

Spectrum sensing is the capability of the CR to allocate the best available, ideal licensed spectrum to the secondary users (SUs) keeping in mind their quality of service (QoS) without disturbing the primary or licensed users. One of the most challenging tasks in CR systems is spectrum sensing as it requires high accuracy and low complexity for dynamic spectrum access. From Figure 6.5, it is clear that in spectrum sensing, the spectrum sensing performance metric is measured between selectivity and sensitivity, which are expressed in terms of levels of detection and false alarm probability. Higher detection probability ensures better primary user (*PU*) protection and lower false alarm probability ensures more chances of channel utilization by secondary user (*SU*). A false alarm probability of 10% and a detection probability of 90% have been regarded as the target requirements for all the sensing algorithms.

To enhance the spectrum sensing efficiency, a hybrid spectrum sensing technique is implemented as shown in Figure 6.6.

To sense the spectrum efficiently, five spectrum sensing techniques are clubbed together to create a hybrid spectrum sensing technique. This hybrid method is summarized as follows: hybrid spectrum sensing method is based upon centralized coordination concept in which an infrastructure deployment is preferred for the CR users. Once CR detects the presence of a primary transmitter, it informs CR controller, which can be a wired immobile device. The CR controller informs all the CR users in its range using broadcast control message.

Centralized schemes can be further classified according to their level of cooperation.

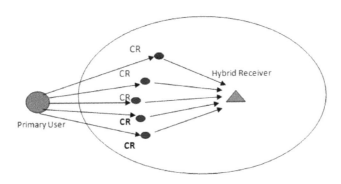

FIGURE 6.6
Hybrid spectrum sensing structure in a cognitive radio network. (From Khobragade, A.S. and Raut, R.D., *Inter. J. Elect. Comput. Eng. (IJECE)*, 7, 2683–2695, 2017.)

Partially Cooperative Scheme: In this scheme, the network nodes cooperate only in sensing the channel. CR users independently detect the channel and inform the CR controller about it.

Totally Cooperative Schemes: In this scheme, the nodes cooperate in relaying each other's information in addition to cooperative sensing channel.

The hybrid system consists of five spectrum sensing methods:

1. Match filter detector
2. Energy detector
3. GLRT
4. Robust estimator correlator
5. Temperature-based detector

Energy detector and match filter detector methods are already discussed earlier. The remaining methods are explained in brief.

6.5.1 Generalized Likelihood Ratio Test

The model of Generalized Likelihood Ratio Test (GLRT) is shown in Figure 6.7, and spectrum sensing shows reduction in probability of false alarm versus the probability of detection for different input energy levels and hence the energy shall also be a parameter in deciding the effectiveness of cognitive systems besides their optimization techniques for energy efficiency. For multipath fading, the GLRT is considered as the best method in demodulator domain for channel estimation and data detection. It is regarded as blind because the channel fading coefficients are unknown to both the receiver and transmitter. The GLRT is considered the best decision maker for this work. The GLRT criterion is based on the following two assumptions.

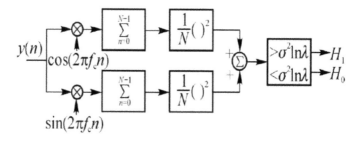

FIGURE 6.7
Principle of the GLRT detector. (From Khobragade, A.S. and Raut, R.D., *Inter. J. Elect. Comput. Eng. (IJECE)*, 7, 2683–2695, 2017.)

1. Both transmitter and receiver know the memory order of the channel v.
2. None of them has the knowledge about channel fading λ.

The received signal can be expressed in the form

$$y(n) = \theta\, A\cos\left(2\pi\, fc\, n + \varphi\right) + w(n);$$

$$n = 0,1,\dots, N\ -1,\ \theta = 0 \text{ or } 1$$

where A is the amplitude, φ is the phase, f_c is the carrier frequency, and $\theta = 0$ or 1 represents the primary signal is absent or present, respectively; the test statistic is as follows

$$T(y) = P\left(y(n);\ A, \varphi, H_1\right) / P\left(y(n); H_0\right) = \lambda \qquad (6.6)$$

where $p(y(n); H_1)$ and $p(y(n); H_0)$ are the PDF of the received signal under each hypothesis.

However, A and φ represent the maximum likelihood estimation (MLE) of the amplitude and the phase, respectively. The likelihood function under H_0 in the condition of the unknown parameter A and φ is

$$P\left(y(n); H_0\right) = 1 / \left(2\Pi\sigma^2{}_w\right)^{N/2} \exp\left[-1/2\sigma^2{}_w \Sigma^{N-1}_{n=0} y^2(n)\right] \qquad (6.7)$$

If the signal is present, the likelihood function under H_1 can be expressed as

$$P\left(y(n); A, \varphi, H_1\right) = 1 / \left(2\Pi\sigma^2{}_w\right)^{N/2} \exp\left(-1 / \left(2\Pi\sigma^2{}_w\right)\right)$$

$$\sum_{0}^{N-1}\left(y(n) - A\cos\left(2\Pi fcn + \varphi\right)\right)^2 \qquad (6.8)$$

Putting above two equations in test statistic, we get

$$T_G(y) = 1/N\left(\left(\sum_{n=0}^{N-1} y(n)\cos 2\Pi fcn\right)^2 + \left[\sum_{n=0}^{N-1} y(n)\sin 2\Pi fcn\right]^2\right) >< H_1 H_0 \lambda G \qquad (6.9)$$

6.5.2 Robust Estimator Correlator

The more we know a priori how a system behaves, the better we can optimize the performance. Exploiting prior knowledge of a system has proven to be a source of performance improvement in many scenarios. However, if this knowledge is relied upon without taking into consideration the possibility of errors or deviations, the result can be degraded. In this method, robust designs are provided for the signals to be emitted from a multiantenna array by taking into account uncertainties in the system parameters. By characterizing the possible parameter uncertainties or errors, the design can combat against severe performance degradation. In general design problems, there exists an objective function $f(Dv, Sp)$ where design variables Dv should be chosen such that $f(Dv, Sp)$, for given system parameters Sp, is optimized. Moreover, a set of constraints over Dv, described by the set Dv, is also satisfied. Mathematically, 1 min $Dv f(Dv, Sp)$ s.t. $Dv \in Dv$. (1.1).

The performance degradation occurs when a physical phenomenon is described by an erroneous model with (possibly) uncertain system parameters. Time variations are also important causes of errors. Even if one estimates system characteristics perfectly, these characteristics will change over time, which makes the estimates unreliable. As a result, signal processing or control schemes are devised by incorporating such uncertainties. In practice, some levels of loss are inevitable, where the performance of a system is evaluated under real-world assumptions, that is using real parameters that essentially fall away from prior knowledge based on which the whole processes and designs have taken place. Let the MIMO base-band system model between the primary user base station and the secondary user, corresponding to the kth symbol transmission, be represented as,

$$y(k) = Hx(k) + \eta(k) \tag{6.10}$$

where $y(k) \in C^{Nr*1}$, and $x(k) \in C^{Nr*1}$ are the received and transmitted temporally independent and identically distributed (IID) zero-mean Gaussian signal vectors, respectively, with the Gaussian signal covariance matrix defined as $R_s \in C^{Nr*Nr}$

$$R_s = E\{s(k)s^H(k)\} = H.E\{x(k)x^H(k)\}H^H \tag{6.11}$$

6.5.3 Robust Test Static Detector

The test statistic corresponding to RTSD for spectrum sensing in MIMO cognitive radio scenarios can be equivalently given as,

$$T_{RTSD} = \sum_{k-1}^{K} y^H(k)R^{-1}{}_{\eta}y(k) - f_{RTSD*} \tag{6.12}$$

where f_{RTSD*} denotes the optimal value of the objective function for the optimization framework.

6.5.4 Robust Estimator Correlator Detector

The optimization framework in the test statistic can be equivalently formulated as

$$T_{RECD} = \sum\nolimits_{k-1}^{K} y^H(k) R^{-1}{}_\eta y(k) - f^*{}_{RECD}$$

where $f^*{}_{RECD}$ denote the optimal value of the objective function. The above test statistic yields a robust decision rule for primary user detection in MIMO cognitive radio networks.

6.5.5 Interference-Based Detection

Unlike the primary receiver detection, the basic idea behind the interference temperature management is to set up an upper interference limit for given frequency band in specific geographic location such that the CR users are not allowed to cause harmful interference while using the specific band in specific area. Typically, CR user transmitters control their interference by regulating their transmission power (out of band emissions) based on their locations with respect to primary users. This method basically concentrates on measuring interference at the receiver. The operating principle of this method is like a UWB technology where the CR users are allowed to coexist and transmit simultaneously with primary users using low transmit power that is restricted by the interference temperature level so as not to cause harmful interference to primary users. The interference temperature model is shown in Figure 6.8.

Here, CR users do not perform spectrum sensing for spectrum opportunities and can transmit right way with specified preset power mask. However, the CR users cannot transmit their data with higher power even if the licensed system is completely idle, since they are not allowed to transmit with higher than the preset power to limit the interference at primary users. It is noted that the CR users in this method are required to know the location and corresponding upper level of allowed transmit power levels. Otherwise they will interfere with the primary user transmissions.

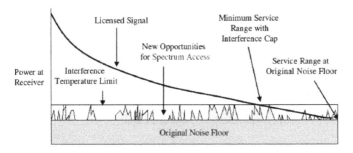

FIGURE 6.8
Interference temperature model. (From Khobragade, A.S. and Raut, R.D., *Inter. J. Elect. Comput. Eng. (IJECE)*, 7, 2683–2695, 2017.)

In hybrid spectrum sensing technique, all five detectors are arranged in a cooperative manner.

1. Match filter detector
2. Energy detector
3. GLRT
4. Robust estimator correlator
5. Temperature-based detector

To perform the hybrid spectrum sensing method, the system will take the input as the received signal and will test this received signal on all the five detector methods, then compare the outputs with the threshold value, and then finally output is declared whether the band is free or occupied.

The algorithm is as follows:

Step 1: Take input signal.

Step 2: Give input to the match filter method.

Step 3: Record the output.

Step 4: Repeat step 2 and step 3 for all the other four methods.

Step 5: If number of true results recorded divide by 5 is greater than 0.66.

Then return True.

Else return False.

To detect whether band is vacant or occupied, the frequency band is sensed by all the above methods, and the result of each method is then sent to the function block where the results are compared with the fixed threshold value and a final output is determined. The flow of the system is as follows.

In Figure 6.9, the system is provided with a spectrum band as the input to check whether it is free or occupied. The input module sends the information of the band to the spectrum sensing module. In this module, the system will test the individual bands of the spectrum. The same input is provided to all

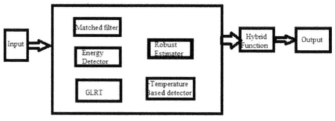

Spectrum Sensing Techniques

FIGURE 6.9
Flow of the system. (From Khobragade, A.S. and Raut, R.D., *Inter. J. Elect. Comput. Eng. (IJECE)*, 7, 2683–2695, 2017.)

the five methods of the spectrum sensing block. That is, if Channel 2 is to be checked, then all the sub-blocks will receive Channel 2 credentials only. Once all the sub-blocks process the channel status and provide the output, then this output is sent to the functional block. In the functional block, the output is checked. If the output of majority of sub-blocks takes either of the side that the channel is free or not, then the final output will be displayed.

For example, let us assume that the outputs from the sub-blocks are as follows: $M = 1, E = 1, G = 1, R = 0, T = 1$. From the output, we observed that four detectors tell us that the band is free out of the five detectors taken. This result is then sent to the hybrid function block. In the function module, the system will check for the majority; if the majority of the output is true and is greater than the threshold of 66%, then the channel is free with a chance of 80%, since four methods convey that the channel is free out of five methods. Therefore, the final output indicates that the band is vacant. In this way, multiple bands can be tested.

In order to compare the performance of different spectrum sensing methods, two basic tasks are considered:

1. To determine which bands are unoccupied.
2. Best method that can be determined for assigning secondary users to the unoccupied spectrum without interfering with the primary users. Using MATLAB simulation of above detectors, it can be experimentally checked that which band is unoccupied band. Here, different spectrum sensing methods are amalgamated and compared based on a variety of performance metrics. Such performance metrics include detection accuracy, complexity, robustness, flexibility of design choices, RF spectrum used, and execution time of each system. In the below table, spectrum sensing algorithms such as energy detector, matched filtering, GLRT, robust estimator correlator, and a hybrid technique are analyzed and compared.

The best spectrum decision method depends on the application. Some variables to be considered are:

1. Expected SNR values.
2. Computational and implementation complexity.
3. Required reliability (express in terms of probability of detection, probability of false alarm).
4. Amount of information that the receiver knows about the primary user's transmitted signal.
5. Execution time.
6. Flexibility in primary user's transmitted signal.

The choice of a particular method purely depends on the application and environment in which it will be used. From Figure 6.10, it can be seen that the

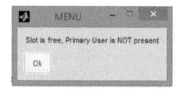

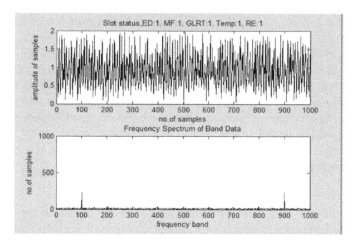

FIGURE 6.10
Output of hybrid system in terms of slot status. (From Khobragade, A.S. and Raut, R.D., *Inter. J. Elect. Comput. Eng. (IJECE)*, 7, 2683–2695, 2017.)

output from the sub-blocks are as follows: $M = 1$, $E = 1$, $G = 1$, $R = 1$, $T = 1$. Hence, the analysis concludes that all five detectors say that the band is free. This result is then sent to the function block. In the function module, the system will check for the majority, if the majority of the output is true and is greater than the threshold of 66%, then the channel is free with a chance of 100%. Since five methods say that band is vacant, the final output is displayed as band is free. In this way, multiple channels can be tested. After a certain time interval, if we again check the same channel, we observe that the result is different, which depends on the status of primary user.

Table 6.1 presents the comparison of all methods.

From Figure 6.11, it can be observed that the output from the sub-blocks are as follows: $M = 1$, $E = 0$, $G = 0$, $R = 0$, $T = 0$. Hence, we observed that only one detector specifies that the band is free out of the five detectors. This result is then sent to the function block. In the function module, the system will check for the majority; if the majority of the output is true and is greater than the threshold of 66%, then the band is free with a chance of 20%, since only one method says that channel is free out of five methods. Therefore, the final output is that the slot is occupied by the primary user. In this way, multiple channels can be tested.

TABLE 6.1

Comparison of Different Methods

	Energy Detector	Matched-Filter	GLRT	Robust Estimator	Interference Based	Hybrid Method
Detection accuracy	Good performance at high rate and poor performance at low SNR, high noise could show false detection	Best performance at all SNRs (if receiver has sufficient knowledge of transmitter) and poor performance if insufficient data is known	Best performance at all SNRs (it needs no knowledge of noise variance)	Good performance at all SNRs	Measuring interference at the receiver	Good performance at all SNRs
Complexity	Low computations and implementation complexity, requires higher no. of samples to converge	High complexity (requires a dedicated receiver for each primary signal class) requires fewest no. of samples to converge	Optimal when the sample size goes to infinity+	Robust designs (the signals to be emitted from a multiantenna array)	Medium complexity (transmitters control totally depends on transmission power)	Medium complexity, if channel symbol positions are unknown, must estimate positions
Robustness	Does not require any prior information of transmitted signal	Requires perfect transmitter information at receiver. It needs a dedicated receiver for every type of primary user	Received signal samples at different antennas/ receivers are usually	Prior knowledge of a system has proven to be a source of performance improvement	Regulating their transmission power	Requires perfect channel symbol positions for dedicated receives
Design parameters	Difficult to choose decision threshold	Transmitter characteristics can be chosen to improve accuracy	Statistical covariance matrix of the primary signals is required	The design performance depends on erroneous model with uncertain system parameters	Specific geographic location	Fixed threshold, cooperative concept and dedicated channel symbols are used to improve accuracy

(Continued)

TABLE 6.1 (*Continued*)
Comparison of Different Methods

	Energy Detector	Matched-Filter	GLRT	Robust Estimator	Interference Based	Hybrid Method
Types of RF spectrum	Does not work for spread spectrum signals	RF spectrum	TV band	MIMO baseband system model	UWB technology	Spectrum diversity concept is used
Execution time	Sensing time taken to achieve a given probability of detection may be high	Matched filter detection needs less detection time because it requires only 0/1 samples to meet a given probability of detection constraint	Execution time depends on propagation channel dispersive time	Time variations are also important causes of errors	Execution totally depends on restriction by the interference temperature level	It performs sensing at periodic time intervals as sensed information become fast due to factors like mobility, channel impairments, etc. Hence, execution time is very small (no delay)

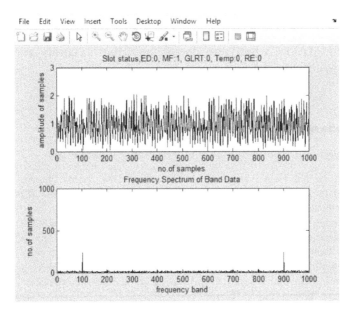

FIGURE 6.11
Output of hybrid system in terms of slot status. (From Khobragade, A.S. and Raut, R.D., *Inter. J. Elect. Comput. Eng. (IJECE)*, 7, 2683–2695, 2017.)

6.6 Conclusions

Spectrum utilization can be improved significantly by allowing secondary users to utilize licensed bands when the primary users (*PU*) are absent. Cognitive radio (*CR*), as an agile radio technology, has been proposed to promote the efficient use of the spectrum.

The spectrum sensing method appears as a crucial need to achieve satisfactory results in terms of efficient use of available spectrum and limited interference with the licensed primary users. To overcome these challenges, hybrid spectrum sensing method is deliberated in this chapter.

The implementation of the hybrid spectrum sensing algorithm for cognitive radio network requires the involvement and interaction of many advanced techniques, which includes cooperative spectrum sensing, interference management, cognitive radio reconfiguration management, and distributed spectrum sensing communications. Further for efficient utilization of limited radio frequency spectrum, the method used should identify the interference. Agility improvement of cooperative networks reduces detection time compared to uncoordinated networks. It also provides fast, reliable operation with proper identification of the system interferences. The throughput of secondary nodes can be improved by increasing the spatial diversity and spectrum diversity. The simulations done using MATLAB Software demonstrate the feasibility and performance of hybrid method.

References

1. J. Mitola, III, G.Q. Maguire, Jr. 1999. Cognitive radio: Making software radios more personal, *Personal Communications, IEEE [see also IEEE Wireless Communications]*, 6: 13.
2. C. Cordeiro, K. Challapali, D. Birru, N.S. Shankar. 2005. IEEE 802.22: The first worldwide wireless standard based on cognitive radios. *IEEE International Symposium on Dynamic Spectrum Access Networks*, 328–337.
3. I.F. Akyildiz, W.-Y. Lee, M.C. Vuran, S. Mohanty. 2006. Next generation/dynamic spectrum access/cognitive radio wireless networks: A survey. *Computer Network*, 50: 2127–2159.
4. F. F. Digham, M-S Alouini, M.K. Simon. 2007. On the energy detection of unknown signals over fading channels. *Proceeding IEEE Transactions on Communications*, 55: 21–24.
5. A. Rama Krishna, A.S.N. Chakravarthy, A.S.C.S. Sastry. 2016. A hybrid cryptographic system for secured device to device communication. *International Journal of Electrical and Computer Engineering (IJECE)*, 6: 2962–2970.
6. Z. Bhutto, J.-H. Park, W. Yoon. 2006. Characterizing Multi-radio energy consumption in cellular/Wi-Fi smartphones. *International Journal of Electrical and Computer Engineering (IJECE)*, 6(6): 2920–2930.
7. J. Lunden, V. Koivunen, A. Huttunen, H.V. Poor. 2007. Spectrum sensing in cognitive radios based on multiple cyclic frequencies. In *Proceeding 2nd International Conference Cognitive Radio Oriented Wireless Network Communication* (CrownCom), Orlando, FL.
8. Y. Ganesan, G. Li. 2007. Cooperative spectrum sensing in cognitive radio *Part I: Two user networks. IEEE Transactions Wireless Communication*, 6: 2204–2213.
9. G. Ganesan, Y.G. Li. 2007. Cooperative spectrum sensing in cognitive radio *Part II: Multiuser networks. IEEE Transactions Wireless Communication*, 6: 2214–2222.

7

Codec Design

7.1 Introduction

Wi-Fi (wireless fidelity) is one of the most popular wireless communication standards in the market. Wi-Fi technology was almost solely used to wirelessly connect laptop computers to the Internet via local area networks. However, this technology has certain limitations. Security and interference are the main issues with current Wi-Fi standards, as well as its inability to reliably stream high-definition audio and video. The problem with Wi-Fi access is that hot spots cover a very small area, so coverage is sparse. To overcome these problems, we need a new technology that would provide:

1. High speed broadband service
2. Wireless rather than wired access, so it would be a lot less expensive than cable or DSL and much easier to extend to suburban and rural areas
3. Broad coverage like the cell phone network instead of small Wi-Fi hotspots

This system is actually coming into being right now, and it is called as WiMAX. *WiMAX* is short form for *Worldwide Interoperability for Microwave Access*, and it also goes by the name *IEEE 802.16* [1]. WiMAX has the potential to provide broadband Internet access. The result is that many people have given up their "land lines." Similarly, WiMAX is expected to replace cable and DSL services, providing universal Internet access just about anywhere you go. WiMAX will also be as painless as Wi-Fi, turning your computer on will automatically connect you to the closest available WiMAX antenna.

The IEEE 802.16, that is, the WiMAX system has the following specifications:

- Range: 30-mile (50-km) radius from base station
- Speed: 70 Mbps
- Line of sight not needed between user and base station
- Frequency bands: 2–11 GHz and 10–66 GHz (licensed unlicensed bands)

TABLE 7.1

Comparison of WiMAX and Wi-Fi

	WiMax	Wi-Fi
Standard	802.16	802.11
Range	30-mile radius	4–6-mile radius
Speed	70 Mbps	2–54 Mbps
Channel Size	1.75–20 MHz	At least 20 MHz
Frequency Bands	2–11 and 10–66 GHz	2.4–5 GHz
Networking employed	MAN	LAN
Multiplexing	TDM/FDM	TDM
Mobility	Vehicular	Pedestrian

Table 7.1 gives the comparison of WiMAX and Wi-Fi.

In this chapter, a detailed discussion on the optimization approach for the codec design in a WiMAX transceiver system is deliberated. The WiMAX transceiver system and a novel approach of using turbo code in concatenation with Reed–Solomon (RS) code in codec design of WiMAX system with OQPSK modulation technique is discussed in detail and implemented in WiMAX system. Furthermore, implementation of the four models, namely,

1. RS in concatenation with convolution codes (CCs) (present system)
2. RS in concatenation with turbo codes
3. Codec employing only turbo codes
4. The OQPSK model is discussed

At the end, a comparative study of the bit error rate (BER) is presented on the basis of the simulation results.

7.2 Transceiver System in WiMAX

The receiver and antenna could be a small box or they could be built into a laptop the way Wi-Fi access is today. A single WiMAX tower can provide coverage to a very large area, as big as 3,000 square miles (~8,000 square km).

The block diagram of a WiMAX transmitter is shown in Figure 7.1. The data in a WiMAX transmitter is processed through the various blocks as explained below.

7.2.1 Data Source

The information bits must be randomized before the transmission. The randomization process is used to minimize the possibility of transmissions of

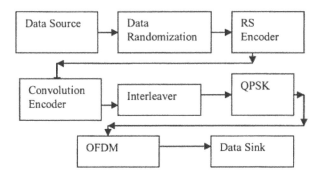

FIGURE 7.1
Block diagram of a WiMAX transmitter.

nonmodulated subcarriers. The process of randomization is performed on each burst of data on the downlink and uplink and on each allocation of a data block (subchannels on the frequency domain and orthogonal frequency division multiplexing (OFDM) symbols on the time domain). Randomized data is passed to the encoder. The technology has claimed to provide shared data rates up to 70 Mbps, which is enough to support more than 60 businesses with Tl-type connectivity.

7.2.2 Encoder

The channel encoder has the task of reducing the BER to ensure secure communications. Moreover, concatenation has proved to yield better results wherein encoding techniques with contrast features have been employed to build a concatenated encoder. Here, the encoding g process consists of a concatenation of an outer RS code and an inner CC as a forward error correction (FEC) scheme. This means that first data passes in block format through the RS encoder, and then it goes across the convolution encoder. It is a flexible coding process due to the puncturing of the signal and allows different coding rates. The last part of the encoder is a process of interleaving to avoid long error bursts.

7.2.3 Interleaver

Data interleaving is generally used to scatter error bursts and thus reduce the error concentration to be corrected with the purpose of increasing the efficiency of FEC by spreading burst errors introduced by the transmission channel over a longer time. Interleaving is normally implemented by using a two-dimensional array buffer, such that the data enters the buffer in rows, which specify the number of interleaving levels, and then it is read out in columns. The result is that a burst of errors in the channel after interleaving becomes in few scarcely spaced single symbol errors, which are more easily correctable. This data is superimposed on high-frequency carrier in the modulator.

7.2.4 QPSK Modulator

Phase shift keying is a modulation process whereby the input signal, a binary PCM waveform, shifts the output waveform to one of a fixed number of states.

The general analytic expression for PSK is

$$s_i(t) = (2E/T)^{\wedge} 1/2 * \cos\left[\omega_0 t + \varphi_i(t)\right]$$

$0 < t < T$, $i = I, ..., M$, where the phase term $\varphi_i(t)$ will have M discrete values, typically given by $\varphi_i(t) = 2\pi i/M$, $i = 1, ..., M$. E is the energy, T is the time duration, and $0 < t < T$. WiMAX system employs a QPSK modulator. A QPSK-modulated carrier undergoes four distinct changes in phase that are represented as symbols and can take on the values of and $\pi/4$, $3\pi/4$, $5\pi/4$, and $7\pi/4$. Each symbol represents two binary bits of data. The QPSK modulated data is frequency multiplexed using OFDM.

7.2.5 Orthogonal Frequency Division Multiplexing

The word orthogonal indicates that there is a precise mathematical relationship between the frequencies of the carriers in the system. The available transmission bandwidth (BW) is divided by orthogonal overlapping of narrow band subchannel. To obtain high spectral efficiency, the frequency response of the subcarriers are overlapping and orthogonal thus the name OFDM. OFDM is resistant to multipath effects that hamper deployment of wireless broadband systems. Multipath is the phenomenon where a radio signal arrives at a receiver via two or more paths. These multiple occurrences of the signal interfere with each other, causing degradation in the overall received signal. Causes of multipath include reflections off objects and refraction. The key advantage of OFDM over single carrier modulation schemes is the ability to deliver higher BW efficiency. This allows the standard to deliver a higher throughput over the link, which is very desirable and cost-efficient.

7.2.6 WiMAX Receiver

The receiver basically performs the reverse operation as the transmitter. In addition, channel estimation is also carried out at receiver. OFDM symbol is composed of data, subcarriers, and some guard bands. Thus, a process to separate all these subcarriers is needed. First, the guard bands are removed, and then a disassembling is performed to obtain pilots, data, and trainings. Once the data has been demapped, it enters the decoder block. The final stage of receiver processing is the decoder. The decoder accepts the sequence of bits and, in accordance with the encoding method that was used, attempts to reproduce the information originally generated by the source. The deinterleaver rearranges the bits from each burst in the correct way by ordering them consecutively as before the interleaving process.

7.3 Novel Approach for Codec Design in WiMAX

As shown in Figure 7.1, codec in the existing WiMAX system uses an RS encoder in concatenation with a normal convolution encoder. The properties of RS codes make them suitable to applications where errors occur in bursts. RS error correction is a coding scheme that works by first constructing a polynomial from the data symbols to be transmitted and then sending an over sampled version of the polynomial instead of the original symbols themselves. The error correction ability of any RS code is determined by $(n - k)$, the measure of redundancy in the block. If the location of the erroneous symbols is not known in advance, then an RS code can correct up to t symbols, where t can be expressed as $t = (n-k)/2$, n is the number of bits after encoding, and k is the number of data bits before encoding. After the RS encoding process, the data bits are further encoded by a binary convolutional encoder.

A convolutional encoder converts the entire input stream into length n codewords independent of the length k. The development of convolutional codes is based mostly on physical construction techniques [2]. The evaluation and the nature of the design of convolutional codes depend less on an algebraic manipulation and more on construction of the encoder. A codec design has been proposed wherein a turbo encoder is used in concatenation with an RS encoder. Turbo coding represents a new and very powerful error control technique, which has started to have a significant impact in the late 1990s, allowing communication very close to the channel capacity. They were introduced in 1993 by Berrou, Glavieux, and Thitimajshima [3]. The powerful error correction capability of turbo codes was recognized and accepted for almost all types of channels, leading to increased data rates and improved quality of service. The introduction of turbo codes made signaling possible at power efficiencies close to the Shannon's limit of 1.59 dB. In principle, turbo code is the parallel concatenation of two or more systematic codes. Turbo decoder uses iterative decoding. The turbo code decoder is based on a modified Viterbi algorithm that incorporates reliability values to improve decoding performance. The turbo decoder consists of M elementary decoders, one for each encoder in turbo encoding part. Each elementary decoder uses the soft decision Viterbi decoding to produce a soft decision for each received bit.

After an iteration of the decoding process, every elementary decoder shares its soft decision output with the other $(M-1)$ elementary decoders. The MATLAB Simulink implementation of a WiMAX system using RS in concatenation with turbo is described in section to follow [4,5]. The use of offset quadrature phase-shift keying (OQPSK) has been proposed, in place of the QPSK modulator–demodulator of the existing WiMAX system. Here, we are using an offset of $\pi/4$. The $\pi/4$ shifted QPSK modulation is a quadrature phase shift keying technique that can be demodulated in a coherent or noncoherent fashion. In $\pi/4$ QPSK, the maximum phase shift is limited to ±135 degrees. Hence, the band limited $\pi/4$ QPSK signal preserves the constant envelope property better than band-limited QPSK. Taking four values of the phase (two bits) at a time to construct a QPSK symbol can allow the phase of the signal to jump by as much as 1800 at a time.

When the signal is low-pass filtered (as in a typical transmitter), these phase shifts result in large amplitude fluctuations, an undesirable quality in communication systems. By offsetting the timing of the odd and even bits by one bit-period, or half a symbol period as shown below, the in-phase and quadrature components will never change at the same time. This will limit the phase shift to no more than 900 at a time. This yields much lower amplitude fluctuations than nonoffset QPSK and is sometimes preferred in practice. A simulation model of the WiMAX using OQPSK is prepared in MATLAB (Figure 7.2).

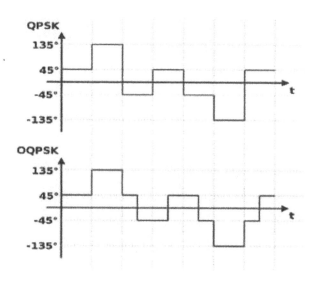

FIGURE 7.2
Comparative phase shifts in QPSK and OQPSK.

7.4 Implementation of the Novel Codec Design in MATLAB

For modeling the WiMAX system, the following major blocks are to be designed and implemented.

7.4.1 Data Source

In the simulation model, instead of performing a randomization process, an integer source that produces random sequences of integers is used. The number of integers that are generated is specified to be frame-based and is calculated from the packet size required in each situation.

7.4.2 RS Encoder

As specified in the standard, the RS encoding is derived from a systematic RS (n = 255, k = 239, t = 8) code using a Galois field specified as GF (2^8). The primitive and generator polynomials used for the systematic code are expressed as follows:

Primitive polynomial:

$$p(x) = x^8 + x^4 + x^3 + x^2 + 1.$$

Generator polynomial:

$$g(x) = \left(x + \beta^0\right)\left(x + \beta^1\right)\left(x + \beta^2\right)\ldots\left(x + \beta^{2t-1}\right)$$

The primitive polynomial is the one used to construct the symbol field, and it can also be named as field generator polynomial. The code generator polynomial is used to calculate parity symbols and has the form specified as before, where β is the primitive element of the Galois field over which the input message is defined. To make the RS code flexible, that is, to allow for variable block sizes and variable correction capabilities, it is shortened and punctured. When a block is shortened to k bytes, 239 k zero bytes are added as a prefix, and, after the encoding process, the 239 k encoded zero bytes are discarded. Once the process of shortening has been done, the number of symbols going in and out of the RS encoder change along with the number of symbols that can be corrected (t). With the puncturing, only the first 2t of the total 16 parity bytes shall be employed. Figure 7.3 shows the RS encoding, shortening, and puncturing process, and Figure 7.4 shows the MATLAB simulation of RS encoder–decoder for WiMAX.

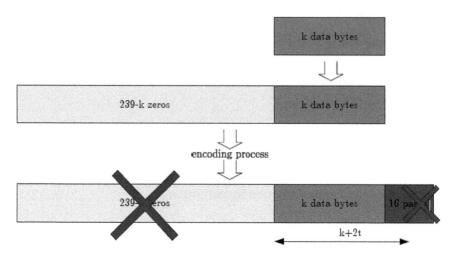

FIGURE 7.3
Process of shortening and puncturing of the RS code.

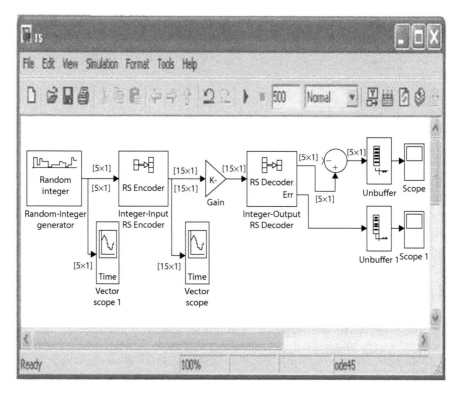

FIGURE 7.4
MATLAB simulation of RS encoder–decoder for WiMAX.

7.4.3 Convolution Encoder

IEEE WiMax 802.16d standard yields a complex convolutional encoder with a constraint length of 7 (see Figure 7.5).

The generator polynomials for this encoder are $g0 = 171$ oct and $gl = 133$ oct. The encoder can easily be implemented in hardware shift registers. The first step is to represent the input bit string as a polynomial. Any sequence of 0s and 1s can be represented as a binary number or a polynomial. The convolutional encoder for WiMax ($g0 = 171$ oct and $gl = 133$ oct) can be represented as follows:

$$g0 = 1 + D + D^2 + D^3 + D^6$$

$$g1 = 1 + D^2 + D^3 + D^5 + D^6$$

The convolutional encoder basically multiplies the generator polynomials by the input bit string as follows:

$$A(x) = g0(x) * I(x) = a\ b\ c...\ g$$

$$B(x) = g1(x) * I(x) = P\ Q\ R\ ...\ V$$

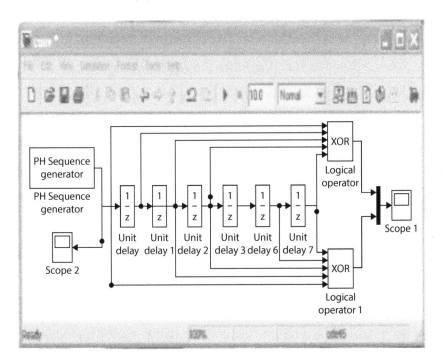

FIGURE 7.5
MATLAB simulation of convolution encoder.

Interleaving the two outputs from the convolutional encoder yields

$$E(x) = a \, P \, bQcR... \, gV$$

which can also be written as

$$E(x) = (a0 \, b0 \, c0 \, ... \, g0) + (0P0Q0R...0V) = A \, x(2) + x * B \, (x2)$$

Therefore,

$$E(x) = A(x2) + x * B(x2)$$

where

$$A(x2) = g0(x2) * I(x2)$$

$$B(x2) = g1(x2) * I(x2)$$

Therefore,

$$E(x) = g0(x2) * I(x2) + x * gI(x2) * I(x2)$$
$$= I(x2) \, (g0(x2) + x * gI(x2))$$
$$= I(x2) * G(x)$$

where

$$G(x) = (g0(x2) + x * gI(x2))$$

Thus,

$$G(x) = 1 + x + x^2 + x^4 + x^5 + x^6 + x^7 + x^{11} + x^{12} + x^{13}$$

7.4.4 Turbo Encoder

It consists of two convolutional encoders. The outputs of the turbo encoder are the information sequence, together with the corresponding parity sequence produced by first encoder and the parity sequence produced by the second encoder block, the input to second encoder is through interleaver, which scrambles the data bit sequence. Simulation model of turbo encoder–decoder is shown in Figure 7.6.

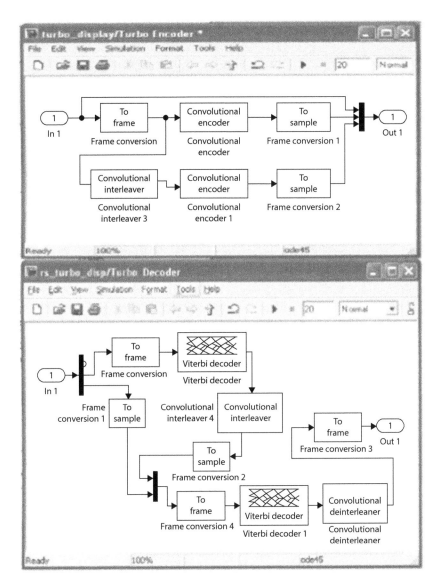

FIGURE 7.6
Simulink model of turbo encoder–decoder.

7.4.5 Turbo Decoder

Turbo decoder shown in Figure 7.6 uses iterative decoding. The turbo code decoder is based on a modified Viterbi algorithm that incorporates reliability values to improve decoding performance. The turbo decoder consists of elementary decoders—one for each encoder in turbo encoding part. Each elementary decoder uses the soft decision Viterbi decoding to produce a soft decision for each received bit. After an iteration of the decoding process, every elementary decoder shares its soft decision output with the other M–1 elementary decoders.

7.4.6 Channel

The AWGN Channel block adds white Gaussian noise to a real or complex input signal. Each of the major blocks mentioned above have individual subblocks, which are configured to meet the WiMAX specifications (after scaling, keeping in mind the mathematical constraints of modeling a real-time system). Figure 7.7 shows the entire WiMAX transceiver system using RS in concatenation with turbo codes.

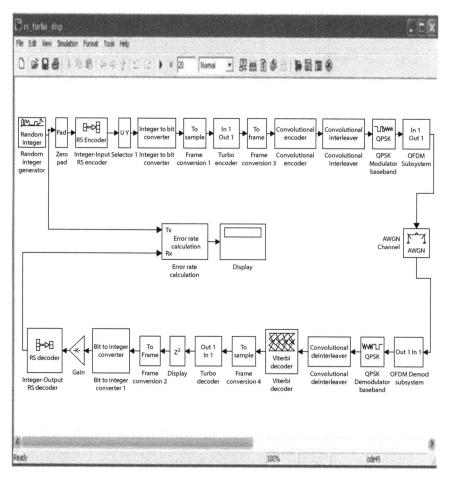

FIGURE 7.7
WiMAX transceiver system employing a codec using RS + turbo.

Furthermore, the simulation models prepared for four different systems are as follows:

1. WiMAX using RS + convolutional codec and OQPSK
2. Existing WiMAX system (RS + convolutional) and QPSK
3. WiMAX using RS + turbo codec and QPSK
4. WiMAX using turbo codec and QPSK

Each of them was observed for BER at different values of S/N (signal-to-noise ratio). Graphs of S/N versus FER (frame error rate) were plotted and are shown in Figure 7.8.

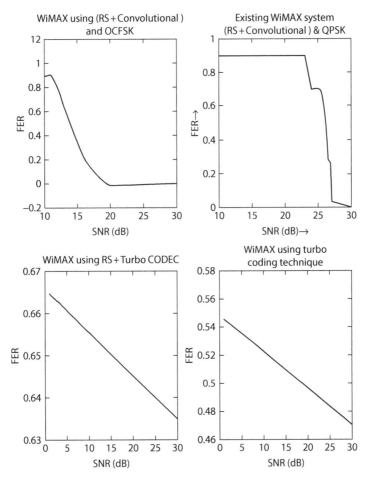

FIGURE 7.8
Plots of S/N vs. frame error rate (FER) for different models of WiMAX system.

From the WiMAX simulation results, following conclusions can be drawn.

1. As seen from Figure 9.8, the turbo concatenation with RS gives desired performance of decreasing the BER with increasing S/N.
2. Using only turbo codes for the codec design, works perfect as far as simulation is concerned. But real-time systems will introduce burst errors. To reduce burst errors, RS coding is a must.
3. Moreover, it was observed that every system requires different signal power for the receiver to interpret the data. The turbo-concatenated system works fine with low power signals as compared to only RS codec system.
4. Also, the OQPSK system can respond to much low-level signals in comparison with the existing QPSK systems and produce a much lower BER.

To observe the results in the form of bits on the computer screen and to plot the graphs, the working frequencies were scaled down. The model works equally well at high frequencies and was tested for few sample data at high frequencies. The problem faced was that the simulation time taken increases to hours due to the computational time required in finding the checksum bits that are derived from input bit streams. High frequency is high data bit rate, that is, more number of bits in a given frame. Every checksum bit is derived with a combinational and sequential logic utilizing the input as the input bit stream. The first model is of a simple nonadaptive digital communication system (DCOM) wherein the codec does not adapt to the available BW. It is evident from Figure 7.9 that as the BW increases, the BER also increases.

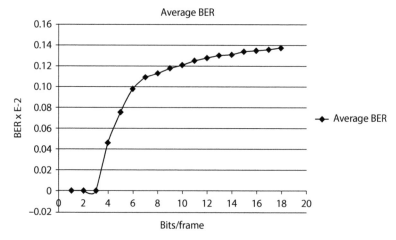

FIGURE 7.9
BER on Y-axis vs. bits per frame on X-axis for nonadaptive codec.

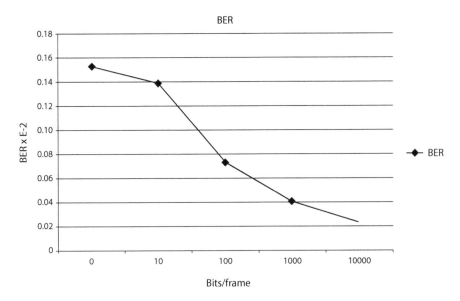

FIGURE 7.10
BER on Y-axis vs. bits per frame on X-axis for adaptive codec.

The second model employs an algorithm to adapt to the increasing BW and utilizing this BW to make the data more secure, that is, to reduce the BER. The graph in Figure 7.10 shows that the adaptive codec reduces the BER with increase in frame size that has resulted from increase in BW.

For performance optimization, one should first think of the parameter to be optimized. Look into the related parameters that improve along with the chosen parameter. The most important of all the contradicting ends should be taken care of; their value should remain within the tolerable limits.

7.5 Optimal Codec Design for Mobile Communication System

Optimization process needs a case study, including simulation to observe real-time results. With the evolution of coding techniques like turbo, low-density parity check code (LDPC), convolutional LDPC, RS codes, and so on, to choose a specific technique and its specific configuration to optimize the system performance is a critical task. Moreover, FPGA implementation of these coders makes the design flexible and software defined. Heading toward the cognitive technology, wherein BW allocation is dynamic and one has to configure the entire transceiver system to work over the available BW [6].

The traditional codes cannot be used for mobile communication. The major difficulty of traditional codes is that, in an effort to approach the theoretical limit for Shannon's channel capacity, there is a need to increase the codeword length of a linear block code. Otherwise, one has to constraint the length of a convolutional code, which, in turn, causes the computational complexity of a decoder to increase exponentially. Random codes are known to achieve Shannon limit performance as k gets large, but at the price of a prohibitively complex decoding algorithm. Ultimately, a point is reached where complexity of decoder is so high that it becomes physically unrealizable. Also, wireless channels are mostly affected by burst errors.

However, the existing codes are effectively suited for correcting random errors, but not burst errors. So, the goal is to design a channel coding scheme for wireless communication that can operate at low S/N and perform reliably. The way to combat the problem is to use concatenated coding, where two (or more) constituent codes are used in serial or in parallel. Concatenated coding schemes are proposed as a method for achieving large coding gains by combining two or more relatively simple building block or component codes. The mobile communication system model with optimal codec is shown in Figure 7.11 [7].

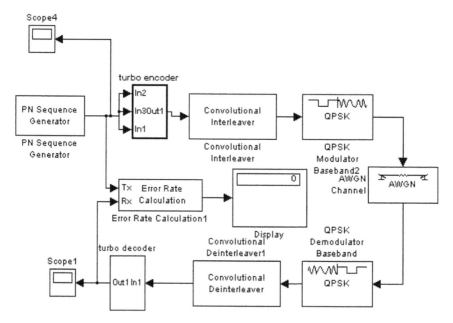

FIGURE 7.11
Simulink model for mobile communication system.

For modeling the WiMAX system, the major blocks required for design and implementation are already discussed in Section 7.4. For the spectrum sensing cognitive radios, once a BW is available and if the available BW is wide enough, then there are two options:

- To transmit bulk data at higher data rate, for example, real-time applications like mobile services
- To transmit small packets of data with high accuracy, required in emergency services (time-bound emergency information should take care of data reduction to minimum possible size so as to utilize low data rates where BER is low)

However, the emphasis here is to improve the BER performance of the system. The first step in modeling is to prepare a simple DCOM model to test the encoder decoder design. This was done using MATLAB Simulink. After testing the functionality of the codec in the DCOM system, a gate-level model of the encoder was prepared, noting the fact that MATLAB Simulink to HDL converter supports only the behavioral-level blocks. These steps are to be carried out to test the hardware implementation possibility of the codec. The cognitive cycle begins with radio scene analysis, scanning for spectrum holes. Once the BW is available, in the available BW, the transmit frequency is decided. So, an M-File for BW selection was written. Now, the input signal in a DCOM is a digital data (binary bits), obtained from an analog source. Common example of analog source is a speech signal, which varies from 300 to 3300 Hz.

This will decide how many digital data bits are present over the observation time. In the model, data bits are randomly generated to represent random speech signal. The bit generation frequency also changes, considering the variations in human speech. We are not concerned with the amplitude variations. Adaption is with regards to frequency, and it should be shown that it is dependent on channel BW and independent of the frequencies in input data. Therefore, the model has been worked out for varying frequencies of input data, and it is observed that it performs equally well over the entire specified range of frequencies of input data. Thus, the model now takes the shape with three parameters as user defined, namely:

1. Input data start frequency
2. Input data end frequency
3. Observation time

A simple DCOM for commercial communication working over the available BW was modeled with rate ½ convolution encoder. This was termed as a nonadaptive system, wherein it was found that as BW increases, the BER also increases. To make the system adaptive, that is as BW increases, one

TABLE 7.2

Bits per Frame vs. Computational Time and Corresponding BER

Sr. No.	Bits per Frame	Computation Time	BER
1	5	3 min	0.1340 E-2
2	7	7 min	0.1184 E-2
3	10	15 min 10 sec	1.1071 E-2
4	15	1 hour 36 min	0.0750 E-2

should be able to send data with higher accuracy, the codec algorithm is made efficient with additional bits and additional sequential and combinational logic. This definitely increases the computational time but drastically reduces the BER. High-speed processors should take care of this increase in computational time. Typical values of computation time required for varying number of bits per frame is quoted in Table 7.2.

The convolution encoder in MATLAB was HDL coded, and the code can be downloaded in FPGA. The entire work flow can be briefly depicted in the block diagram shown in Figure 7.12.

The simulation results for the mobile communication system models are shown in Figure 7.13. It is plot of BER vs. S/N wherein decoder iterations is set as a parameter. BER reduces with increase in decoder iterations.

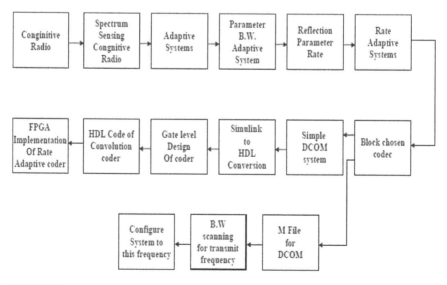

FIGURE 7.12
Work flow.

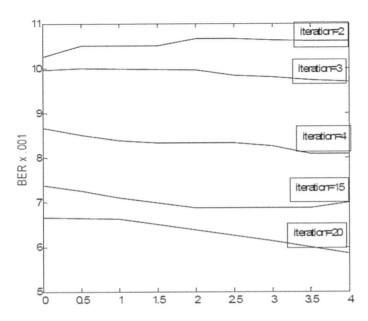

FIGURE 7.13
BER Vs. E_b/N_o, for mobile communication system.

7.6 Modeling and Simulation of Spectrum Sensing Smart Codec for Cognitive Radio

The first step in modeling a simple DCOM model is to test the encoder decoder design. This is done using MATLAB Simulink. The work model is shown in Figure 7.14. After testing the functionality of the codec in the DCOM system, a gate-level model of the encoder is prepared, noting the fact that MATLAB Simulink to HDL converter supports only the behavioral-level blocks, this is depicted in Figure 7.15. The cognitive cycle begins with radio scene analysis, scanning for spectrum holes. Once the BW is available, in the available BW, the transmit frequency is decided.

So, an M-File for BW selection is necessary. Now the input signal in a DCOM is a digital data (binary bits) obtained from an analog source. Common example of an analog source is a speech signal, which varies from 300 to 3300 Hz. This will decide how many digital data bits are present over the observation time. In the model, data bits are randomly generated to represent random speech signal. The bit generation frequency also changes, considering the variations in human speech. The amplitude variations here are neglected. Adaption is with regards to frequency, and it should be shown that it is dependent on channel BW and independent of the frequencies of

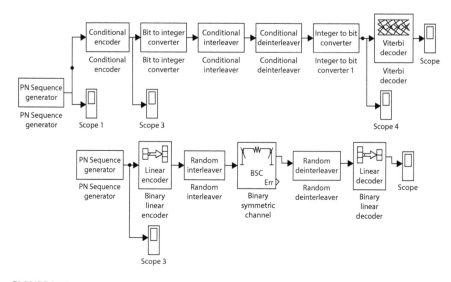

FIGURE 7.14
Simulation model of simple digital communication system, prepared in MATLAB Simulink.

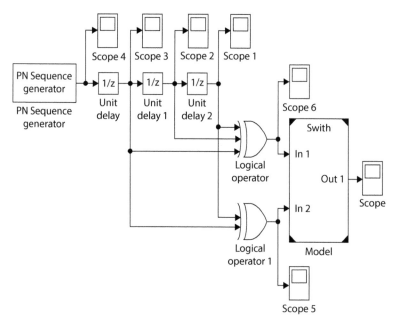

FIGURE 7.15
Gate-level design of the convolution encoder for Simulink to HDL conversion.

input data. The model has been analyzed for varying frequencies of input data and performs equally well over the entire specified range of frequencies of input data. Thus, the model now takes the shape with three parameters as user defined, namely:

1. Input data start frequency
2. Input data end frequency
3. Observation time

A simple DCOM for commercial communication working over the available BW requires a rate of ½ convolution encoder. This is termed as a nonadaptive system, wherein the BER increases with increase in bandwidth. The system is made adaptive, that is, as BW increases, one should be able to send data with higher accuracy, by employing the codec algorithm that can add additional bits and additional sequential and combinational logic. This definitely increases the computational time but drastically reduces the BER. High-speed processors take care of this increase in computational time. Typical values of computation time required for varying number of bits per frame is quoted in Table 7.2.

The convolution encoder in MATLAB and coded in HDL can be downloaded in FPGA. The input data bits and coded output of the rate ½ encoder is also analyzed.

To observe the results in the form of bits on the computer screen and to plot the graphs, the working frequencies should be scaled down. The model works equally well at high frequencies and can be tested for few sample data at high frequencies. The problem that may be faced during simulation is that the time taken increases to hours due to the computational time required in finding the checksum bits that are derived from input bit streams. High frequency is high data bit rate, that is, more number of bits in a given frame. Every checksum bit is derived with a combinational and sequential logic utilizing the input as the input bit stream.

To summarize, the first model represents a simple non-adaptive DCOM wherein the codec does not adapt to the available BW. In this case, as the BW increases, the BER also increases. The second model is adaptive, that is, the increasing BW can be utilized and the data can be made more secure, that is, to reduce the BER.

In an adaptive codec, the BER reduces with increase in frame size that has resulted from increase in BW.

The simulation output of the FPGA implemented HDL code of convolution encoder is shown in Figure 7.16 [8,9].

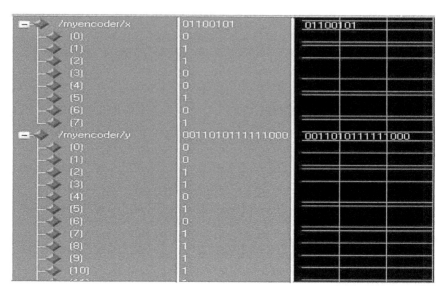

FIGURE 7.16
0–7 input bits and 0–10 output coded bits waveform.

7.7 Conclusions

This chapter mainly focuses on the codec design. Different designs of codec for WiMAX, mobile communication system, are discussed in detail. At the end, design of a spectrum sensing codec is also discussed in detail. Smart Codec uses CCs wherein the codec senses the spectrum available and accordingly modifies the algorithm for coding–decoding. An increase in BW is utilized for reducing the BER. A comparison of both adaptive and nonadaptive codec is discussed. The hardware encoder had been realized by downloading the HDL code into FPGA.

References

1. R.D. Raut, K.D. Kulat. 2008. Novel approach: Codec design for WiMAX system, *IEEE Explore,* 18–20.
2. T. Ohtsuki. 1999. Rate adaptive indoor infrared wireless communication systems using repeated and punctured convolutional codes. *1999 IEEE International Conference on Communications (Cat. No. 99CH36311).* Vol. 1. IEEE, Dresden, Germany.

3. C. Berrou, A. Glavieux, P. Thitmajshima, 1993. Near Shannon limit error-correcting coding and decoding: turbo codes. *Proceeding IEEE International Conference. Communication*, Geneva, Switzerland.

4. Matlab 7 Getting Started Guide, Available: http://www. Mathwaorks.com

5. Simulink Fixed Point 5 User's Guide, Available: http://www. Mathwaorks.com

6. M. Chiani. 2008. Coexistence of ultra-wide band and other wireless systems: The path towards cognitive radio, *UWB Workshop*, Ferrara, Italy.

7. R.D. Raut, K.D. Kulat. 2008. Optimal codec design for mobile communication, TECHNIA. *International Journal of Computing Science & Communication Technologies*, 1(1): 20–24.

8. D.L. Perry. 2002. *VHDL Programming by Example*, 4th edn. TMH, McGraw-Hill, New York.

9. FPGA & CPLD Solutions from Xilinx Inc., Available: http://www.xilinx.com.

8

A Review of Security Issues
in Cognitive Radio

8.1 Introduction

Wireless network technology has witnessed an amazing escalation during the previous few decades. In 1985, the Federal Communications Commission (FCC) issued a mandate authorization defining quite a few portions of the electromagnetic spectrum as "license-exempt." Some bands of the spectrum were permitted to operate devoid of the need of license and this was called as the Industrial, Scienfic, and Medical (ISM) band. This announcement of FCC and the arrival of IEEE 802.11/a/b/g standards have brought a revolution in the wireless realm. The ISM band, being license free, has thus become congested, resulting in augmented intrusion and contention. Although the ISM band is overcrowded, there are several accredited bands of spectrum, which are being underutilized.

Licensed band here refers to a band of spectrum, which is allotted by FCC by the conventional and static allocation procedure of spectrum assignment serving a distinct service or several channels of distinct service. The static allocation, on the other hand, differs from country to country; it has been experiential that the parts of spectrum are not utilized resourcefully. It was Joseph Mitola, who anticipated a novel idea of the opportunistic use of spectrum, which was underutilized. He proposed an idea in the form of a novel device called cognitive radio (CR). CR is the phrase coined by Mitola and Maguire in an article in which they describe it as a radio that understands the milieu in which it finds itself and as a consequence can mold the communication process in line with that understanding. The interconnection of CRs will form a cognitive radio network (CRN).

Simon Haykin defines CR as an intelligent wireless communication system that is aware of its surrounding environments (i.e., outside world) and uses the methodology of understanding by building to learn from the environment and adapt its internal states to statistical variations in the incoming RF stimuli by making corresponding changes in certain operating parameters (e.g., transmit power, carrier frequency, and modulation strategy) in real time. A CR is a device that has four broad inputs, namely,

1. An understanding of the milieu in which it operates
2. An understanding of the communication obligation of the user(s)
3. An understanding of the authoritarian policies that can apply to it
4. An understanding of its own capabilities. One of the foremost concerns in any wireless communication is that of security. This chapter focuses primarily on the security issues associated to the CR.

8.2 CR Basics

CR in a way is a "well-groomed radio." Generally, there are two types of CRs: policy radios and learning radios. Policy radios, by looking at the word "policy," we can conclude that it is a radio that has to go behind some predefined policies that will make a decision concerning its actions. Learning radios on the other hand have a learning engine that permits the radio to become skilled from its ambiance and can organize or reconfigure its state. These radios make the use of an assortment of artificial intelligence (AI) learning algorithms as well.

In CR state of affairs, there will be two categories of users: (i) primary users [PUs], holding a permit of a particular portion of spectrum, and (ii) secondary users [SUs], also called cognitive users who do not hold the license but still can use the segment of spectrum allotted to the PUs in an opportunistic approach. However, this approach should not hinder or hamper the performance of the licensed user. The most important objective of CR is to sense the spectrum and make its opportunistic use. The free spectrum portions are referred to as *"white spaces"* or *"spectrum hole."*

CR technology is a wireless network know-how and is susceptible to all conventional threats of wireless networks in addition to the inimitable threats introduced with the initiation of this innovative area of wireless communication. CR is based on what is called a *software-defined radio (SDR).* The software allows the radio to tune to diverse frequencies, power levels, modulation depending upon its erudition and the milieu in which it operates. The CR is expected to carry out the following four functions:

1. *Spectrum Sensing*: revealing the "white spaces" or the segment of spectrum not in use.

   This must also make sure that no PU (licensed user) is in service at the same time.

2. *Spectrum Management*: choice of the best spectrum hole for broadcast.

3. *Spectrum Sharing*: distribution of spectrum with other probable users.

4. *Spectrum Mobility*: check out of the band when a licensed user is perceived *(spectrum handoff)*.

The functions mentioned above form a cognition cycle that forms the foundation on which the CR will function. In his thesis, Mitola has explained the cognitive cycle consisting of five states, namely, *Observe, Orient, Plan, Decide,* and *Act*. Figure 8.1 shows the cognitive cycle described by Mitola.

Till now, we have talked about basics of CR. The study of any wireless network would be partial unless we explore its topology. As far as the CR topology is concerned, three different architectures come into picture. These are as follows:

1. Infrastructure
2. Ad-hoc
3. Mesh

Infrastructure networks have base station (BS) also called as access points. A device can communicate with other devices within the neighborhood through the base station. The communication amid these devices can be routed via the BS. Ad-hoc topology is formed by devices devoid of the need of a BS. These communicate with each other by establishing links amid themselves, making use of diverse communication protocols available. Mesh topology is an amalgamation of the infrastructure and ad-hoc topologies.

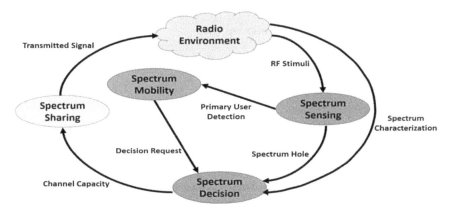

FIGURE 8.1
The cognition cycle.

8.3 Security Threats in CR

A *"security threat"* is described as a probable violation of defense wherein it can be premeditated like an intentional assault or inadvertent owing to an internal failure or malfunctions. An assault is considered burly if it involves a minimal number of adversaries performing minimal operations, but causing utmost harm or loss to the PUs and/or CR.

A thorough examination of defense requirements is defined as follows:

1. *Confidentiality*: confidentiality of the stored and communicated data must be ensured by the system.
2. *Robustness*: the system must defy against assaults and endow with communication services as per service level accord.
3. *Regulatory framework compliance*: the system must adhere to the rules and regulations or policies.
4. *Controlled access to resources*: right of entry to information or resources by unauthorized users must be rejected.
5. *Nonrepudiation*: nondenial of the accountability for any of the actions performed by an entity of the system.
6. *System integrity*: system must guarantee the veracity of its system components.
7. *Data integrity*: system must ensure that the data stored or communicated must not be illegitimately modified.

The CR technology must put into effect the security triad of confidentiality, integrity, and availability (CIA).

The attacks in general follow a layered approach, and as a consequence, the threats are categorized according to diverse protocol layers they aim. The attacks are categorized as follows:

1. Physical layer attacks
2. Link layer attacks (also known as MAC attacks)
3. Network layer attacks
4. Transport layer attacks
5. Application layer attacks

In addition to above, we also converse cross-layer attacks, which are exclusively targeting one particular layer but have an effect on the performance of a different layer.

As described in the diverse classes of attacks, which an assailant can construct is categorized as follows:

1. Dynamic spectrum access attacks
2. Objective function attacks
3. Malicious behavior attacks

8.4 Physical Layer Attacks

It is the bottom layer of the protocol stack and provides an interface to the broadcast medium. It consists of a medium that makes two network devices communicate with each other such as cables, network cards, etc. In CR, the medium is atmosphere. The physical layer determines the bandwidth, channel capacity, bit rate, etc. In CR, the spectrum is accessed as and when it is found untenanted, this makes the physical layer intricate. Figure 8.2 shows the physical layer attacks.

8.4.1 PU Emulation Attack

A primary function of CR is to sense the spectrum that we refer to as spectrum sensing and the spectrum has to be shared in an opportunistic approach. The SU has to evacuate the currently used spectrum as and when it detects an incumbent (PU) signal to avoid the interference. This is referred to as the spectrum hand off. For a fair spectrum sharing, it is obligatory that the CR must be acquainted with the PU signals.

Nodes launching PUEA are of two types:

Greedy nodes, which broadcast counterfeit PU signals forcing all other users to evacuate the band in order to acquire its exclusive use.

Malicious nodes copy PU signals in order to cause denial of service (DoS) attacks. These nodes can lend a hand and can broadcast counterfeit PU signals in more than one band ensuing in hopping of a CRN form band to band, hampering its complete operation.

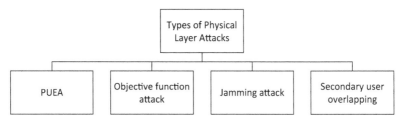

FIGURE 8.2
Physical layer attacks.

In PUEA, the assailant emulates the PU signal to access the resources. The SU will be beneath the notion that a PU is using the spectrum band, and thus, it will evacuate the band.

Now, if the attacker's goal is to augment its share of spectrum, then we refer this as self-centered PUEA. This assault can be conducted concurrently by two attackers by establishing a devoted link amid them. If the attacker's goal is to thwart other genuine users from using the spectrum, then we refer this assault as a malicious PUEA. The PUEA can aim at both types of CRs, the policy radios and learning radios with different severity. Figure 8.3 shows a typical PUEA scenario.

8.4.2 Objective Function Attack

The radio parameters comprise of center frequency, bandwidth, power, modulation type, coding rate, channel access protocol, encryption type, and frame size. The cognitive engine manipulates these parameters to meet one or more objective functions. The objective function attacks are those in which algorithms that exploit the objective functions are attacked. Another name for objective function attacks is "belief-manipulation attacks."

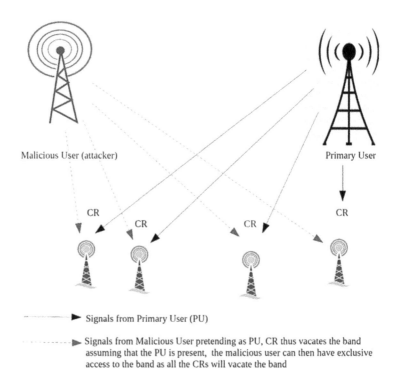

FIGURE 8.3
PUEA scenario.

Intrusion detection systems (IDSs) are used for detecting the objective function attack. The IDS may be a misuse-based method or anomaly-based method. The first method uses the signatures of assault, while the other makes use of the unusual behavior of the system to notice an attack. Thus, the attacks not recognized beforehand can be detected by the later system. It is appealing to mention here that the IDS follows the FCC constriction of not modifying the PU system. IDS operates in two phases: the profiling phase and the detection phase.

8.4.3 Jamming

Analogous to the conventional wireless technology, the CR receiver requires a minimum signal-to-noise ratio (SNR) decode a digital signal. One of the customary assault strategies is to trim down the SNR below a threshold value by transmitting noise over the channel this is also referred to as "receiver jamming."

Now there may be a case wherein the performance of the receiver node is poor, this may or may not be due to jammer. Sometimes this performance may be poor due to natural causes, say network congestion, for example. Jamming is an assault that can be done in the physical and the MAC layers. There are four types of jammers, namely

1. Constant Jammer
2. Deceptive Jammer
3. Random Jammer
4. Reactive Jammer

1. *Constant Jammer*: A constant jammer is the one that is continuously sending out data packets, while doing so it will not wait for channel to be idle and also it has no regard for other users on the channel.
2. *Deceptive Jammer*: A deceptive jammer will send out packets continuously such that other users will switch to receive states and will linger in that state.
3. *Random Jammer*: A random jammer will take breaks amid jamming signals, and it may behave as a constant or a deceptive jammer during the jamming phase.
4. *Reactive Jammer*: A reactive jammer is the one that will sense the channel all the time. It starts transmitting the jamming signals each time it detects a communication on the channel and thus it is complicated to detect a reactive jammer.

"Intentional Jamming" is one of the most basic types of attacks in which the assailant incessantly and purposely transmits data packets on a licensed band, making it unusable for both the primary and SUs. In Primary Receiver

Jamming, an assailant close to the PU will send request for transmission from other SUs. This will turn away the traffic toward the PU, which will in turn generate intrusion to the PU.

8.4.4 Overlapping SU

A geographical region may contain coexisting and overlapping secondary networks. An attacker in one network can transmit signals that may cause harm to the primary and SUs of both networks. The malicious node is not in the control of the users of the network or under the direct control of the station. This is a direct attack on the capability of the CR network for spectrum sensing and sharing of both infrastructure and ad-hoc-based networks. This results in the denial of service attack.

8.5 Link Layer Attacks

Figure 8.4 shows different attacks associated with the link layer. We discuss one by one in brief.

8.5.1 SSDF Attack (Byzantine Attack)

The 802.11 data link layer consists of two sublayers. These are LLC and MAC. MAC supports multiple users on a shared medium within the similar network. The "Spectrum Sensing Data Falsification" (SSDF) attack is also called as *"Byzantine Attack."* The attacker in this attack is the legitimate user of the network and is called *"Byzantine."* This assault is for acquiring the spectrum accessibility and occupancy with the goal of destructing the whole communication system. This assault is investigated from two different scenarios.

1. Distributed CRN
2. Centralized CRN

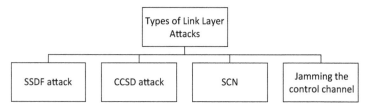

FIGURE 8.4
Link layer attacks.

In a distributed CRN, the spectrum sensing information is based on the observations of the SU itself and on the basis of the observations shared by the SUs. In other words, the spectrum information is shared in collaboration amid the SUs. In a centralized CRN, diverse SUs sense the environment and the spectrum sensing information is sent to a fusion center (FC). The FC is liable for collecting the data and on the basis of this data, the FC provides information regarding frequency bands that are free or busy.

The manipulation of FC data will prevent the user from accessing the unoccupied band or it may permit to use a band that is already in use resulting in intrusion. This attack has more impact in a disseminated CRN in which false information can proliferate rapidly. However, in case of centralized CRN, a smart FC can evaluate the data received by diverse CRs and recognize which CR may be providing counterfeit information.

8.5.2 Control Channel Saturation DoS Attack (CCSD)

In a multi-hop CRN, a channel negotiation is obligatory in a distributed manner. After channel negotiation, the CRs can communicate with each other. Now, when many CRs want to communicate with each other, the channel negotiation becomes difficult as the channel can support only a limited number of data channels. An assailant will consequently try to send MAC control frames, thereby saturating the control channel, which results in poor performance of the network with approximate zero throughput. It has been further observed that this attack has more impact on multihop CRNs as compared to centralized CRNs. This is because of the fact that the MAC control frames are authenticated by the BS and thus forging MAC frames is almost impractical.

8.5.3 Selfish Channel Negotiation (SCN)

A selfish channel in a multihop CRN can decline to forward any data for other hosts. This self-centered behavior will result in conservation of energy and an increase in the throughput. Some other self-centered host alters the MAC behavior of the other CR devices and thus claims the channel at the expense of other hosts. This scenario leads to the severe degradation of the throughput of the entire CRN.

8.5.4 Control Channel Jamming

CR users assist among themselves with the help of control channels. An attacker if attacks a control channel, known as control channel jamming (CCC) attack, the receivers are banned from sensing the valid messages when an attacker injects a strong signal in the control channel. This scenario leads to the denial of service for the users in that particular network.

8.6 Network Layer Attacks

It has been found in literature that much research has been focused on MAC and PHY layers. Network layer helps in routing of data packets from source node to destination node maintaining the quality of service. Routing in CR is a challenge owing to spectrum handoff and dynamic spectrum sensing. As stated earlier, CRs are prone to attacks similar to classic wireless networks. In addition to this, CR is susceptible to attacks that also plague the wireless sensor networks. Figure 8.5 shows the attacks associated with the network layer.

8.6.1 Hello Flood Attack

This attack has been investigated as an attack against a wireless sensor network (WSN) and can be applied to a CR scenario as well due to its routing strategies. In this assault, an attacker first broadcasts a message to all the nodes in a network. This is an advertisement where an attacker offers a high-quality link to a particular destination using high power to persuade the node that the aggressor is a neighbor. As a lot of power is put, the strength of the received signal will be high and the node convinces itself that the attacker is its neighbor even though the attacker is far away from the node. In this way, all nodes will forward data packets to the attacker, assuming that the attacker is their neighbor. However, when the attack is detected, the nodes will lose their data packets and find themselves alone with no neighbor nearby them. Some of the protocols used to exchange information amid neighbors for maintaining topology may also be attacked.

8.6.2 Sinkhole Attack

CR uses multihop routing similar to the WSN. In this type of attack, the attacker will be promoting itself as the best route to a specific destination, thereby enticing the nodes neighboring the attacker. Once the attacker wins

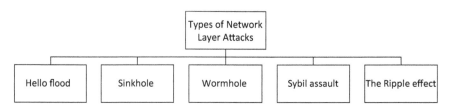

FIGURE 8.5
Network layer attacks.

the trust of its neighbors, the neighbors themselves promote or advertise the attacker's path as the best route. Now, the attacker has the capability to send the received packets directly to the BS using high level power. Once the trust has been established that the attacker's route is the best route, the attacker can begin other attacks, for example, eavesdropping, selective forwarding attack by forwarding data packets from selected nodes, dropping received packets, modifying data packets, and so on.

8.6.3 Worm Hole Attack

It is closely related to the sink hole attack. A wormhole attack is normally perpetrated and administered by two malicious nodes. These nodes understate the distance amid them by relaying the packets along an out-of-bound channel that is unavailable to other nodes. In this attack, the nodes are persuaded by the attacker that they are only one or two hops away from the adversary. However, the nodes will be usually multiple hops from the BS. The attacker thus can take delivery of the packets for forwarding or can capture the packets for eavesdropping. Another appealing scenario here is that the attacker can stop relaying the packets, which would create the separation of the network. Network routing protocols thus have to be implemented and the attacker will be provided additional information, which will help the attacker to perpetrate other potential attacks.

8.6.4 Sybil Attack

Some local entities recognize the other remote entities as informational abstractions (entities) without the physical knowledge of the remote entities. A sybil attacker will create a large number of pseudonymous identities so as to gain a large influence on the network. A system must exhibit the capability to ensure and identify that distinct identities refer to distinct entities. In general, an attacker will pair sybil attack with Byzantine attack or PU emulation attack so that the whole decision-making process gets effected.

8.6.5 Ripple Effect

The CR senses the spectrum and utilizes the portion of spectrum that is free. While doing so it performs spectrum handoff every time it senses the PU so that intrusion is avoided. The ripple effect is similar to Byzantine attack or Primary User Emulation Attack (PUEA). The wrong channel information is shared so that the nodes change their channel. Here, the attacker sends the false information hop by hop so that the network enters a state of confusion.

8.7 Transport Layer Attack—Key Depletion

The protocols used for IEEE 802.11 are prone to key repetition attacks. The transport layer protocols establish cryptographic keys at the beginning of each transport layer session. Due to the generation of large number of keys, there is a possibility that the sessions key that are generated will be repeated. This repetition of keys may result in the breaking of the cipher system.

8.8 Application Layer Attack

This attack is due to the self-propagating behavior of CR and is also known as CR virus. In this attack, a state that is introduced in a CR will cause a behavior that induces the same state in another CR and so on. Thus, the said state will proliferate through all the radios in that particular area. The self-propagating behavior of radio thus can contaminate the entire network. These types of attacks can even spread amid the radios that never had any protocol interaction.

8.9 Cross-Layer Attacks

The cross-layer attacks target the multiple layers and can influence the whole CR cycle. Attacks discussed in previous sections can be combined to form cross-layer attacks. Generally, the cross-layer attacks target one layer but affects the performance of another layer. Figure 8.6 shows cross-layer attacks.

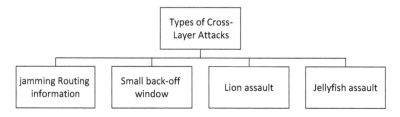

FIGURE 8.6
Cross-layer attacks.

8.9.1 Routing Information Jamming

In this attack, the targeted node performs a spectrum handoff and it stops all ongoing communication, switches to a new spectrum band, identifying the neighboring nodes. The targeted node will not be able to receive the routing information until the spectrum handoff is complete. This is referred to as *deafness*. Till the completion of hand off process, the targeted and its neighboring node will be using the stale route for communication. We know that spectrum hand off is an significant event in CR. There will be some delay while switching from one spectrum band to another spectrum band. This delay during transition is a matter of consideration. This delay allows the attacker to jam the routing information among the nodes. The nodes will therefore follow some stale routes and route the packets on an erroneous route. This attack can be made harsher if the targeted node performs spectrum hand off more frequently.

8.9.2 Small Back-Off Window

This attack is viable against the CR networks using Carrier Sense Multiple Access with Collision Avoidance (CSMA/CA) protocol at the MAC layer. The main motive of this attack is to gain more access to the channel by manipulating the contention protocol parameters (by choosing a very small back-off window) to obtain overall or more frequent access to the channel.

8.9.3 Lion Attack

This attack takes place at the physical layer or the link layer and targets the transport layer. The main intention behind this attack is to obliterate the TCP connection. In this type of attack, the attacker makes use of the PUEA. As we know that the SU, whenever detects a primary signal, has to execute spectrum hand-off. However, the TCP will not be conscious of this hand-off and will continue to send packets as usual without any acknowledgment. As the TCP receives no acknowledgment, it considers that the segment is lost and retransmits the segment, resulting in delays and packet loss. This attack may be more severe if the attacker knows in advance and moves to the channel to which the SU will be hopping to. This state of affairs is analogous to the DoS attack.

8.9.4 Jelly Fish Attack

The jelly fish attack is similar to the lion attack as they both target the TCP. This attack targets the transport layer; however, the attack is performed at the network layer. In this attack, the attacker deliberately reorganizes the packets it receives and forwards. These retransmissions severely degrade the performance. Further, if the malicious node randomly delays the packets, the TCP timers will be worthless, resulting in network congestion. The jelly fish also obeys the data plane and the control plane protocol rules, making it complicated to make a distinction between the attack and congested network.

8.10 SDR Security

SDR security is an additional area of significant importance. SDR security is vital due to its capacity to reconfigure itself. SDR security falls in two main categories:

1. Software-based protection
2. Hardware-based protection

Software-based protection is to defend against the malicious software download, buggy software, installation, and so on. Hardware protection involves the implementation of hardware components. These keep an eye on several other parameters of the SDR.

8.11 Conclusions

It has been experiential that several portions of the licensed spectrum are underutilized, and on the other hand, the user demand for spectrum usage is escalating. The solution proposed is in the form of a hopeful technology called as the CR technology. The CR technology being a wireless technology is defenseless to various attacks including those which also plague the traditional wireless networks as well as WSNs. This technology faces many security threats mainly due to its two attractive characteristics, namely, cognitive capability and reconfigurability. This chapter discusses almost all types of attacks affecting the CR technology. Further, all these attacks are analyzed from a layered approach point of view. The so-called cross-layer attacks and the SDR security threats can have an effect on the overall functioning of the cognitive cycle. At the end, it is concluded that the security of PU is of supreme significance in CR setup.

References

1. S. Haykin. 2005. Cognitive radio: Brain-empowered wireless communications. *IEEE Journal on Selected Areas in Communications* 23: 201–220.
2. L.E. Doyle. 2009. *Essentials of Cognitive Radio*, The Cambridge Wireless Essentials Series. Cambridge University Press, Cambridge, UK.
3. G. Baldini, T. Sturman, A.R. Biswas. 2012. Security aspects in software defined radio and cognitive radio networks: A survey and a way ahead. *IEEE Communications Surveys & Tutorials* 14: 355–379.
4. J. Mitola, 2000. *Cognitive Radio: An Integrated Agent Architecture for Software Defined Radio*, Doctor of Technology, Royal Institute of Technology (KTH), Stockholm, Sweden.
5. A. Fragkiadakis, E. Tragos, I. Askoxylakis, 2012. A survey on security threats and detection techniques in cognitive radio networks. *IEEE Communications Surveys and Tutorials* 15: 428–445.
6. T. Clancy, N. Goergen, 2008. Security in cognitive radio networks: Threats and mitigation. In *Third International Conference on Cognitive Radio Oriented Wireless Networks and Communications* (CrownCom) (2008), pp. 1–8.
7. A. Attar, H. Tang, A.V. Vasilakos, F. Richard Yu, V.C.M. Leung, 2012. A survey of security challenges in cognitive radio networks: Solutions and future research directions. *Proceedings of the IEEE*, 100(12): 3172–3186.
8. C.N. Mathur, K.P. Subbalakshmi, 2007. Security issues in cognitive radio networks. In *Cognitive Networks: Towards Self Aware Networks*, Q.H. Mahmoud (Ed.) pp. 272–290, John Wiley & Sons.
9. R. Chen, J.M. Park, J.M. Reed, 2008. Defense against primary user emulation attacks. *IEEE Journal on Selected Areas in Communications*, 26(1): 25–37.
10. A. Rawat, P. Anand, H. Chen, P. Varshney, 2010. Countering byzantine attacks in cognitive radio networks, acoustics speech and signal processing (ICASSP). In *IEEE International Conference*, pp. 3098–3101.
11. F.K. Jondral, 2005. Software-defined radio—basic and evolution to cognitive radio. *EURASIP Journal on Wireless Communications and Networking*, 3: 275–283.
12. R. Chen, J.M. Park, 2006. Ensuring trustworthy spectrum sensing in cognitive radio networks. In *IEEE Workshop on Networking Technologies for Software Defined Radio (SDR'06)* pp. 110–119.
13. D. Hlavacek, J. Morris Chang, 2014. A layered approach to cognitive radio network security: A survey. *Computer Network*, 75: 414–436.
14. W. El-Hajj, H. Safa, M. Guizani, 2011. Survey of security issues in cognitive radio networks. *Journal of Internet Technology*, 12(2).
15. S. Madbushi, R. Raut, M.S.S. Rukmini, 2015. Security issues in cognitive radio: A review, *Proceedings of the International Conference on Micro-Electronics, Electromagnetics and Telecommunications* (ICMEET) 2015, Lecture Notes in Electrical Engineering (LNEE), Chapter No 12, pp. 121–134, December 18–19, GITAM University, Visakhapatnam, India.

9

Chaotic Communication Approach to Combat Primary User Emulation Attack in Cognitive Radio Networks

9.1 Introduction

In the field of wireless communications, chaotic communication has also been a field of interest. The reason behind this interest is low power consumption, low complexity in design of hardware, robustness in multipath fading environments, low probability of interception, and resistance to jamming. Besides this, chaotic communication provides better security, overcomes the physical constraints faced by wireless systems, and thus provides a better performance. Chaotic communications are based on the chaos theory that describes behavior of nonlinear systems. Such systems are highly sensitive to initial conditions. In literature, chaotic signal detection is divided in two classes: detecting signals contaminated by chaotic signals and detecting signals contaminated by random noise. Chaotic signals are harder to identify, aperiodic and unstable. These signals have low power spectrum density and utilize larger bandwidth. There are numerous features of chaotic signals that make them attractive for use in wireless communications. It is theoretically proved that the Lyapunov exponents remain unaltered if a chaotic signal is utilized. A chaotic system is described by Equation 9.1 shown as follows.

$$x' = A(x) + g(x) \tag{9.1}$$

where $A(x)$ is the linear part and $g(x)$ is the nonlinear part of the system. The chaotic system discussed here is a Lorenz's chaotic system. Its dynamic states are represented by the set of the following three equations as:

$$dx/dt = a(y - x) \tag{9.2}$$

$$dy/dt = x(b - z) - y \tag{9.3}$$

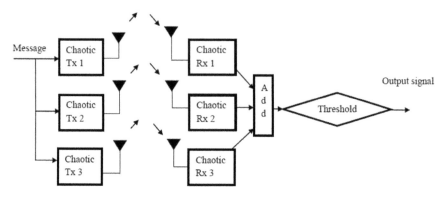

FIGURE 9.1
A three coupled wireless communication system. (From Madbushi, S. et al., *Inter. J. Intell. Eng. Sys.*, 11, 57–67, 2018.)

$$dz/dt = xy - cz \qquad (9.4)$$

where x, y, and z are the dynamic states, a, b, and c are constants greater than zero. The analysis of one coupled, two coupled, and three coupled chaotic systems shows that the three coupled chaotic system's performance is the best. A three coupled system is shown in Figure 9.1.

In chaotic test signal approach, the test signal is masked at first, encoded and decoded using three coupled Lorenz chaotic system, and the BER analysis is done to identify the channel state.

9.2 Tagging Scheme of Test Signal

The message signal is transmitted considering the Binary Phase Shift Keying (BPSK) as the modulation scheme. The code length of the message signal used is 1000 bits (it can be any number of bits, say, 100, 500, 1000, 2000).

Appending the tag bits is a simple procedure wherein the tag bits are inserted in the original message signal in a particular fashion. The procedure of appending the tag bits is shown in Figure 9.2, and the process of chaotic encryption and decryption is explained later. Suppose that there are 20 bits (message bits) and tag bits have to be appended, the tag bit is appended after 1, 5, 9, and 13th bit (odd position) and 2, 6, 10, and 14th bit (even position). The tag bit is denoted by 1. Thus, bit 1 is appended after bit number 13 and after bit number 14. Or, in other words, bit 1 is inserted on both sides of bit number 14. In this way, tag bits can be appended to any number of message bits. Note that the procedure continues and tag bits are appended till the end of message bits.

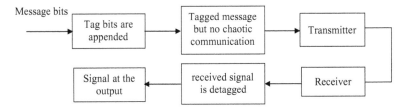

FIGURE 9.2
Inserting tag bits after bit 13 and bit 14. (From Madbushi, S. et al., *Inter. J. Intell. Eng. Sys.*, 11, 57–67, 2018.)

The tagging procedure can be made more secured by employing certain encryption algorithms or techniques. As already stated for chaotic encryption of the message signal three coupled chaotic system is used as shown in Figure 9.1.

Figures 9.3 and 9.4 represent the block diagrams of the proposed methods without and with chaotic communication using tagging, respectively. Further, the BER analysis is also done.

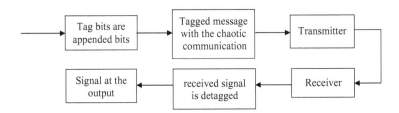

FIGURE 9.3
Model showing tagging without chaotic communication. (From Madbushi, S. et al., *Inter. J. Intell. Eng. Sys.*, 11, 57–67, 2018.)

20 bits - 0 1 0 1 0 1 0 1 0 1 0 1 0 1 0 1 0 1 0 1 (0s are in odd positions and 1s are at even positions)

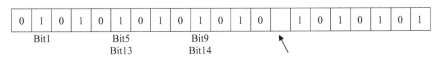

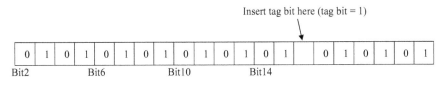

FIGURE 9.4
Model showing tagging with chaotic communication. (From Madbushi, S. et al., *Inter. J. Intell. Eng. Sys.*, 11, 57–67, 2018.)

The new message (original message plus tag bits) is then masked by the chaotic state and transmitted. The encryption and decryption process at the transmitter and receiver are explained as follows:

Encryption (master):

$$x = Ax + g(x, v) + Lz_x$$

where

$$v = x1 + M \qquad (9.5)$$

Decryption (slave):

$$y = Ay + g(y,v) + Lz_y \qquad (9.6)$$

where $x \in R^n$; $y \in R^n$ are the state vectors; Ax and Ay denote the linear part; and $g(x, v)$ and $g(y, v)$ denote the nonlinear part of the system. The controller gain of the system is denoted by L, and the coupling strength between master and slave systems is denoted by K, $(K > 0)$; and z_x and z_y are the feedback signals.

$$x = \begin{bmatrix} x_1 & x_2 & x_3 \end{bmatrix}^T, \; y = \begin{bmatrix} y_1 & y_2 & y_3 \end{bmatrix}^T,$$

$$A = \begin{bmatrix} -a & a & 0 \\ c & -(1+K) & 0 \\ 0 & 0 & -b \end{bmatrix}$$

$$g(x,v) = \begin{bmatrix} 0 & -vx_3 & vx_2 \end{bmatrix}^T$$

$$g(y,v) = \begin{bmatrix} 0 & -vy_3 & vy_2 \end{bmatrix}^T$$

$$L = \begin{bmatrix} l_1 \\ l_2 \\ l_3 \end{bmatrix}, z_x = M \text{ and } z_y = (v - y_1)$$

Here, $l_1 = -a$, $l_2 = a + c$, and $l_3 = 0$.
The synchronization error and its error dynamic are defined as:

$$e = x - y = \begin{bmatrix} e_1 & e_2 & e_3 \end{bmatrix}^T \qquad (9.7)$$

$$\dot{e} = \dot{x} - \dot{y} = Ae + g(x,v) - g(y,v) + L(z_x - z_y) \qquad (9.8)$$

The focal intend is to design L, the controller gain, such that the input message M can be received at the receiver. The message R that will be recovered is

$$R = v - y_1 = (x_1 + M) - y_1 \qquad (9.9)$$

Here, as x_1 equals y_1, $R = M$.

The chaotic system parameters used for the analysis are as follows:

$A = [0.1\ 0.2\ 0.3]$
$M = [0.2\ 0.1\ 0.1]$
$L = [0.15\ 0.25\ 0.35]$
$x_1 = [0.22\ 0.19\ 0.43]$
$z_x = [0.3\ 0.6\ 0.4]$

Figure 9.5 shows the signal of 10 bits in length transmitted and received using tagging and chaotic communication. It can be observed from Figure 9.5 that the original sequence and the received sequence are almost the same and the tagged sequence and the received signal appear as noise.

The chaotic constants must be same at the transmitter and the receiver side. It is difficult to recover the original signal unless the values of the chaotic constants are known, and thus they must be known to both transmitter and receiver for proper recovery of the signal.

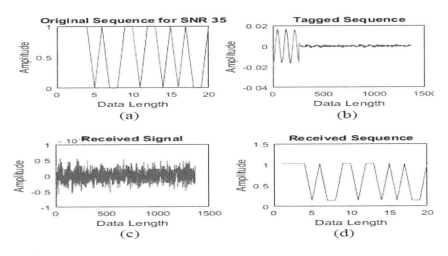

FIGURE 9.5
Signal transmitted and received using chaotic communication. (From Madbushi, S. et al., *Inter. J. Intell. Eng. Sys.*, 11, 57–67, 2018.)

TABLE 9.1

Comparison of BER with Tagging and Tagging with Chaotic Communication Using BPSK

Code Length in Bits	BER with Only Tagging	BER with Tagging and Chaotic Communication	% BER Improved
1000	5.42×10^{-4}	1.62×10^{-4}	71%
2000	2.69×10^{-4}	8.2×10^{-5}	70%
3000	1.75×10^{-4}	5.65×10^{-5}	69%
4000	1.29×10^{-4}	4.25×10^{-5}	69%

Source: Madbushi, S. et al., *Inter. J. Intell. Eng. Sys.*, 11, 57–67, 2018.

Table 9.1 shows the comparison of bit error rate (BER) with only tagging and BER with tagging and chaotic communication using BPSK.

It can be seen from Table 9.1 that the BER is improved with tagging and chaotic communication as compared to only tagging.

Figures 9.6 and 9.7 show the BER versus signal-to-noise ratio (BER vs. SNR) curve for code length equal to 1000 bits using BPSK.

Table 9.1 also shows the percentage improvement for code length equal to 2000, 3000, and 4000 bits, respectively. Table 9.2 shows the comparison of BER with only tagging and BER with tagging and chaotic communication using quadrature phase shift keying (QPSK).

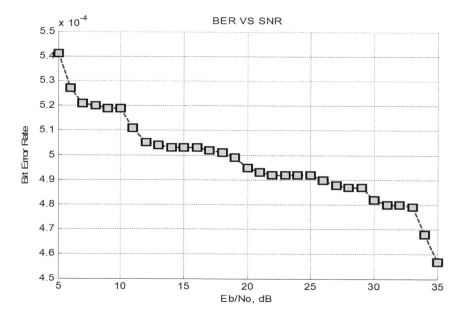

FIGURE 9.6

BER vs. SNR curve with tagging (code length = 1000 bits, BPSK). (From Madbushi, S. et al., *Inter. J. Intell. Eng. Sys.*, 11, 57–67, 2018.)

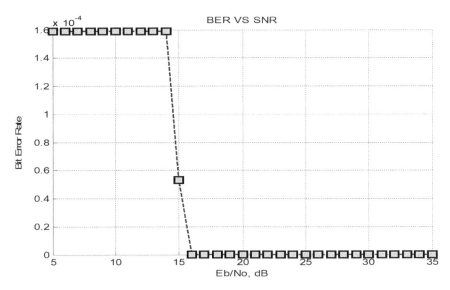

FIGURE 9.7
BER vs. SNR curve with tagging and chaotic communication (code length = 1000 bits, BPSK).
(From Madbushi, S. et al., *Inter. J. Intell. Eng. Sys.*, 11, 57–67, 2018.)

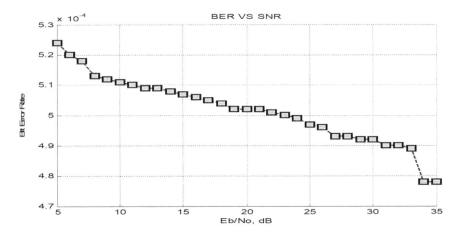

FIGURE 9.8
BER vs. SNR curve with tagging (code length = 1000 bits, QPSK). (From Madbushi, S. et al.,
Inter. J. Intell. Eng. Sys., 11, 57–67, 2018.)

It also shows the percentage improvement for code length equal to
2000, 3000, and 4000 bits respectively using QPSK as modulation scheme.
Figures 9.8 and 9.9 show the BER versus SNR curve for code length equal to
1000 bits using QPSK.

It can be seen from Tables 9.1 and 9.2 that there is an overall improvement of around 69% in the BER of the received signal when tagging with

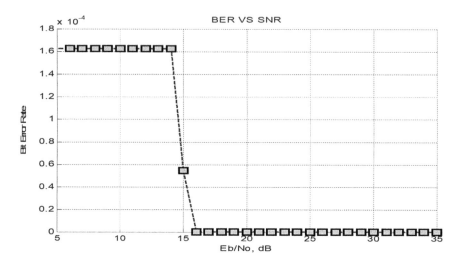

FIGURE 9.9
BER vs. SNR curve with tagging and chaotic communication (code length = 1000 bits, QPSK). (From Madbushi, S. et al., *Inter. J. Intell. Eng. Sys.*, 11, 57–67, 2018.)

TABLE 9.2

Comparison of BER with Tagging and Tagging with Chaotic Communication Using QPSK

Code Length in Bits	BER with Only Tagging	BER with Tagging and Chaotic Communication	% BER Improved
1000	5.26×10^{-4}	1.6×10^{-4}	70%
2000	2.57×10^{-4}	8.25×10^{-5}	69%
3000	1.74×10^{-4}	5.7×10^{-5}	68%
4000	1.32×10^{-4}	4.4×10^{-5}	68%

Source: Madbushi, S. et al., *Inter. J. Intell. Eng. Sys.*, 11, 57–67, 2018.

chaotic communication is considered as compared to only tagging using BPSK or QPSK. It may also be observed that the percentage improvement is not affected much even though the code word is changed from 1000 bits, to 2000 bits, 3000, and 4000 bits, respectively. Thus, it can be concluded that BER has improved considerably when the signal is transmitted using tagging and chaotic communication. Moreover, the signal transmitted using chaotic approach appears as noise and thus provides more secure communication over wireless channel. The next section (Part II) explains how the tagging scheme with chaotic communication is used to send the message/test signal over a channel and detect the Primary User Emulation Attack (PUEA).

9.3 The Test Signal Scheme for Identification of PUEA

At this point, from the results of Part I, it was found that by using tagging and chaotic communication a signal is received and identified properly at the receiver. Also, it is a secured form of communication. In PUEA, one has to identify or differentiate between the original incumbent signal and the signal emulated by the malicious secondary user (attacker). Much work on PUEA focuses on employing three methods mainly energy detection, cyclo-stationary feature detection, matched filter detection, and so on.

Instead of identifying whether the signal is of primary user or some malicious secondary user (attacker) by using conventional methods of signal identification, the problem is analyzed from a different angle. Moreover, no prior knowledge of primary user signal is required.

Whenever a spectrum band is free, it can be used by the secondary users as long as the primary user is not using it. Also, the band has to be vacated as soon as the primary user returns back. Now, if the band is free, the attacker may emulate like a primary signal fooling and avoiding other secondary users to use the band.

Consider a network of 20 users for simulation. In simulation, the status of particular node is denoted as safe = 0 or safe = 1. The nodes denoted as safe = 0 are attackers in the network. The transmission of test signal here takes place from a source node to destination node. Any node, irrespective of its status (safe = 0 or safe = 1) can communicate with any other node. When the test signal is transmitted from a source node whose status is safe = 1 to a destination node whose status is either safe = 0 or safe = 1, the test signal is received properly at the receiver (BER < 0.7, threshold). This means that the channel/band is free and not under PUEA.

On the other hand, if the test signal is transmitted from a source node whose status is safe = 0 to a destination node whose status is either safe = 0 or safe = 1, the test signal is not received properly at the receiver (BER > threshold), and it can be concluded that the particular channel/band is under PUEA. The continuous monitoring of channel is done and scanned for predecided patterns, which are stored at the nonattacking secondary user nodes. These patterns are different for different secondary users, and the network router knows about them in advance.

In order to check if the band is free or not, the secondary user sends out its pattern, this pattern is encrypted using a three-level chaotic encoder for security. The chaotic encryption process makes sure that the pattern behaves like a random noise sequence, and does not interfere with patterns of other secondary users. These two properties are fulfilled by selecting different and orthogonal values for the chaotic constants used in the encryption and decryption process.

Once the nonattacking or genuine secondary user transmits a chaotic sequence, it is decoded by the receiver/router. As the knowledge about the encryption constants is already known to the receiver/router, thus the sequence is decoded properly, with minimum to no errors. This ensures that the BER on the receiver side is either 0 or a minimum value. But, if an attacking node tries to access the channel, by transmitting any random sequences, then the receiver/router will not be able to decode this sequence, and thereby the BER value between the known transmitted and unknown received signal will be very high, and the attacker would be marked.

The detection rate accuracy to identify the attack is 97%. For the same code length of 1000 bits and BPSK modulation and two attackers in the network, the detection rate accuracy is 98.889%. And for three and four attackers in the network, the detection accuracy is approximately 99%. Thus, whenever a band is free a test signal is transmitted using tagging and chaotic communication (the procedure is same as explained earlier) by the secondary user willing to use the band.

At the end, it is summarized that if the test signal sent by the secondary user transmitter is properly received at the secondary user receiver (BER less than or equal to a threshold value), then it is confirmed that the band is free and no attack has taken place. And if the BER of the received signal is greater than the set threshold, then it is confirmed that an attacker is trying to emulate the primary user and the network is under PUEA. One interesting thing to note here is that when the test signal is transmitted by using tagging and chaotic communication, even the attacker will not know that a test signal is being transmitted to check the availability of the spectrum. This is because when the test signal is transmitted using chaotic communication, the test signal appears as noise (see tagged sequence and received sequence in Figure 9.5) and hence cannot be recognized by the attacker.

9.4 Conclusions

This chapter has explored a novel chaotic communication-based approach to combat PUEA in cognitive radio environment confirming to requirement of the FCC. The test signal is tagged at first, then encrypted by a three-level Lorentz chaotic attractor and then transmitted to identify the channel occupancy state. The test signal of different secondary users will not interfere with each other as the selected chaotic constants are different, orthogonal, and appear as random noise. There is no need of a helper node and this reduces the cost of the required infrastructure. The performance of the scheme in terms of the BER shows an improvement in attack detection rate considerably.

References

1. A.A. Sharifi, M. Sharifi, M.J.M. Niya. 2016. Secure cooperative spectrum sensing under primary user emulation attack in cognitive radio networks: Attack-aware threshold selection approach. *AEU-International Journal of Electronics and Communications*, 7(2/3/4): 95–104.
2. M. Haghighat, S.M.S. Sadough. 2012. Cooperative spectrum sensing in cognitive radio networks under primary user emulation attacks. In *Proceeding of Sixth International Symposium on Telecommunications* (IST), Tehran, Iran. pp. 148–151, 2012.
3. H.H. Kuo, T.L. Liao, J.S. Lin, J.J. Yan. 2009. A new structure of chaotic secure communication in wireless AWGN channel. In *Proceeding of International Workshop on Chaos-Fractals Theories and Applications* (IWCFTA '09), Shenyang, China. pp. 182–185.
4. A. Riaz, M. Ali. 2008. Chaotic communications, their applications and advantages over traditional methods of communication. In *Proceeding. of 6th International Symposium on Communication Systems, Networks and Digital Signal Processing*, (CNSDSP), Graz, pp. 21–24.
5. J. Gu, S.H. Sohn, J.M. Kim, M. Jin. 2009. Chaotic characteristic based sensing for cognitive radio. In *Proceeding of 5th International Conference on Wireless Communications, Networking and Mobile Computing*, Beijing, pp. 1–4.
6. R. Chen, J.M. Park. 2006. Ensuring trustworthy spectrum sensing in cognitive radio networks, In *Proceeding of IEEE Workshop on Networking Technologies for Software Defined Radio* (SDR'06), pp. 110–119.
7. R. Chen, J.M. Park, J.H. Reed. 2008. Defence against primary user emulation attacks in cognitive radio networks. *IEEE Journal on Selected Areas in Communications*, 26(1): 25–37.
8. L. Zhang, J. Yu, Z. Wu. 2015. Secured chaotic cognitive radio system using advanced encryption standard. In *Proceeding of IEEE 26th Annual International Symposium on Personal, Indoor, and Mobile Radio Communications(PIMRC)*, Hong Kong, pp. 7–11.
9. S. Madbushi, R. Raut, M.S.S. Rukmini. 2018. A novel chaotic communication based test signal approach for identification of primary user emulation attack in cognitive radio networks. *International Journal of Intelligent Engineering & Systems*, 11(2): 57–67.

10

Bilayer Approach to Mitigate Primary User Emulation Attack in Cognitive Radio Networks

10.1 Introduction

Cognitive radios are disposed to numerous categories of assaults; these can be primary user emulation, self-interested network compromise, control channel compromise, etc. These systems are disposed to such kind of assaults for the reason that they have the intrinsic objects of primary users, secondary users, and channel sensing. Once the system is under attack, the rudimentary properties of cognitive radio networks are worsened and the system acts unpredictably.

In primary user emulation attack, the assailant emulates the functionality of a primary user, and blocks the spectrum so that all the sincere secondary users are deprived of service, for the reason that the primary property of a cognitive radio is that it assigns a channel to one secondary user and preserves its allocation until the communication of the secondary user is completed or until the primary user returns back. This safeguards high quality of service to the primary users, and the channel bandwidth is assigned to other secondary users once the current primary communication is completed. Owing to this property, the network communication is optimized and the channel utilization is evenly managed.

Primary user emulation attackers are likely to be present in all cognitive radio environments. These can be present in the form of radios, which are software-defined, or virtual transreceivers in cognitively skillful devices. These assailants monitor the system traffic and accomplish primary emulation attack when the network usage is maximum. The impact of this is enormous, and in communications primarily, high significance communications get interrupted because of it.

A trust-based mechanism for spotting the source of primary user emulation attacks is deliberated in this chapter. Once the assailant nodes are recognized, they are blocked and detached from the system communication process.

Unpretentious secondary nodes have an individual and distinctive trusted look-up table-based request-challenge mechanism. The main strength of the algorithm is in fact that each honest secondary node is preconfigured and has some unique parts when compared to other genuine secondary nodes.

Primary user emulation attackers have a fewer chance of getting through this trust-based system, but in case they do, then a second layer of attack recognition is applied in which a chaotic communication system is realized. This chaotic communication system encrypts a test sequence from the honest transmitters; this test sequence appears as noise on the channel; thus, it cannot be detected by any attacker node. The testing receiver detects this signal, and if it is received within a proper BER range, then the node is marked to be safe. Otherwise, the node is marked to be unsafe or attacker node and is removed from the network.

The two-layered method aids to detect and eliminate imitation assailants from the network in a very effective and enhanced way. The primary user emulation attackers are noticed and removed at a rate of approximately above 99%, and the network communication is re-established with least possible delay, thus allowing the use of the two-layered method in real-world scenarios, without negotiating the quality of service (QoS) of the system.

10.2 Two-Layered Method for Attack Discovery and Elimination

The two-layer procedure can notice and trace the nodes, which take part in primary user emulation attack. The two-layered method is depicted in Figure 10.1. In the figure, it can be witnessed that the honest nodes have two layers of security inserted into them, the first layer deals with a trust-based look-up table (LUT), which stores key–value or key–expression pairs, whereas the second layer is a chaotic communication layer, which safeguards a second-level check on emulation nodes.

An example of LUT stored in two of the nodes is presented in Table 10.1 and is also present with the router/home node/base station node.

The procedure of trust-based attack elimination is portrayed in Figure 10.2.

The procedure for localizing the attacker node can be demonstrated as follows:

1. SU sends a communication appeal to the router or base station or home node (R).
2. The router (R) recognizes the node number of SU from the appeal and replies with a haphazard challenge (C).

3. SU gets this challenge, and might answer in the following two ways:

   a. If the SU is honest, then it will check the LUT and crack the challenge (C) to get the cracked value (Sv), and then send Sv back to the router.

   b. If the SU is an assailant, then it will reply with a haphazard solution (Sr) to the router.

4. The router will crack the challenge (C) locally by the LUT of the wishing SU node and have the solution (Sc) ready for assessment.

5. If the SU is honest, then Sv will match Sc, and the communication will ensue.

6. If SU is an assailant, then Sr will not be equal to Sc, and the node will be acknowledged as an assailant node. The router will block all communications from this node, and the assailant will be detached from the cognitive radio milieu.

FIGURE 10.1
System model.

TABLE 10.1

LUT Stored at Any Two of the Nodes

Node Number	Input Data Range (x)	Output Value (y)
1	<10	$x*2$
1	<30	$x/2$
1	<50	$(x+2)$
1	<250	$1/x$
1	<500	$\sqrt[3]{x}$
1	<1000	$(x+1)/(x^2-x-1)$
1	>=1000	$(x^2+1)/(x^3+x^2+x+1)$
2	<15	$(x+2)$
2	<40	$(x-2)$
2	<65	$x*2$
2	<125	$(x/2)$
2	<450	$(x+5)$
2	<1200	$(x-1)/(x^2-x-1)$
2	>=1200	$(x^2-x-1)/(x^3+x^2+x+1)$

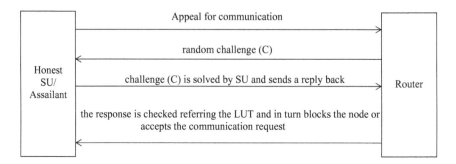

FIGURE 10.2
Trust-based attack removal.

This algorithm can be overcome only beneath two cases.

1. If the assailant knows about the trustworthy LUT of the node
2. If the assailant responds with the correct challenge answer

From the above two cases, it is evident from simulations that case 1 is invalid, as the attacker is usually ad hoc, and will never have the private LUT information of the emulated primary node. But, the second case can happen. In simulations, only 1 out of 1,000,000 times, the attacker can correctly answer the challenge and get access to the communication system. However, this issue can be resolved by increasing the complexity of the LUT key–values or key–expression pairs but it also adds exponentially to the complexity of the overall system, which adds a delay in subsequent communications.

To challenge this situation, another attack detection layer, which is a combination of chaotic communication, tag-based system, and a BER analyzer is designed. In the second layer, predecided patterns are scanned, which are stored at the nonattacking secondary user nodes. These patterns are exclusive for dissimilar secondary users and are known to the network router in advance.

The secondary user will send out its test signal pattern in the second layer; this test signal is encrypted by means of a three-level Lorenz's chaotic attractor encoder for safety. The Lorenz's chaotic attractor is signified by the subsequent three equations

$$\frac{dx}{dt} = \sigma\left(y - x\right)$$

$$\frac{dy}{dt} = x\left(\rho - z\right) - y$$

$$\frac{dz}{dt} = xy - \beta z$$

σ, ρ, and β are the nonzero constants, x, y, and z are the dynamic states. The encrypted test signal configuration acts like a haphazard noise sequence and does not obstruct other secondary user patterns. This is because of the fact that they are orthogonal and have dissimilar values for the chaotic constants used in the encryption and decryption procedure. The nonattacking user transfers a chaotic sequence and is deciphered by the receiver/router. As the receiver/router previously knows the encryption constants, the sequence is decoded correctly, with almost no to least errors so that the BER on the receiver side is either 0 or a nominal value. But, if an offensive node transfers any random sequence to gain admission over the channel, then inappropriate decoding of the sequence will take place at the receiver/router, and the BER value amid the unidentified received signal, and the recognized transmitted signal will be very high. In this way, the assailant would be identified. In simulation process, the BER threshold is 0.7, which guarantees that even if the channel has manifold nonattacking users, then there are minimal false positives detected by the system. The results show an accuracy of more than 99% in detection and localization of the emulation attackers, and thus are very effective in ad-hoc and non-ad-hoc cognitive networking environments. A combination of these two layers guarantees a detection and localization rate of above 95%, which is suited for real-time applications. The delay investigation shows that the scheme can perceive the assailant node in at most two communication sequences, which take less than 1 ms of communication delay per node. There are no in the system. The system will not be overloaded, as there are no complicated, computed-intensive calculations.

Tests under various simulation scenarios are conducted, with a different number of attacker nodes and under varying channel conditions such as the AWGN, Rayleigh, and Rician.

In experiments, the following parameters and values were used in simulation (Table 10.2).

Table 10.3 shows the performance concerning delay and correctness of assailant node detection. The two-layered approach performs very well under various channel conditions; the detection response is quite impressive

TABLE 10.2

Parameters and Values Used in Simulation

Parameter	Value
Number of nodes	10 to 1000
Number of attackers	10% to 20%
FFT Size	64
Carriers	4
Modulation Type	QAM
Chaotic system	3 Level
BER Threshold	0.7

TABLE 10.3

Recognition Rate and Delay Beneath Several Channels and Assailants

Channel Type	Number of Nodes	Number of Attackers	Detection Rate (%)	Mean Delay (ms)
AWGN	10	2	99.95	0.1
AWGN	20	6	99.96	0.3
AWGN	30	10	99.96	0.4
AWGN	40	18	99.97	0.45
AWGN	50	22	99.97	0.55
AWGN	100	35	99.97	0.65
AWGN	500	200	99.98	0.75
AWGN	1000	350	99.98	0.88
AWGN	10000	2500	99.99	0.95
Rayleigh	10	3	99.94	0.15
Rayleigh	20	8	99.95	0.25
Rayleigh	30	12	99.96	0.35
Rayleigh	40	20	99.96	0.5
Rayleigh	50	21	99.97	0.6
Rayleigh	100	30	99.97	0.7
Rayleigh	500	180	99.98	0.76
Rayleigh	1000	320	99.98	0.9
Rayleigh	10000	2300	99.99	0.94
Rician	10	2	99.92	0.18
Rician	20	6	99.93	0.28
Rician	30	9	99.95	0.37
Rician	40	18	99.95	0.49
Rician	50	22	99.96	0.56
Rician	100	25	99.96	0.67
Rician	500	160	99.97	0.71
Rician	1000	280	99.97	0.82
Rician	10000	2400	99.98	0.91

with the system detecting nearly 99% of the attacks, with a delay of fewer than 1 ms for each of the channels.

The overall system performance can be portrayed by the graphs shown in Figures 10.3 and 10.4. Figure 10.3 shows plot of the detection rate versus the number of nodes in AWGN, Rayleigh, and Rician channels. Figure 10.4 shows the plot of delay in detection versus the number of nodes in AWGN, Rayleigh, and Rician channels.

The delay performance of the system starts growing linearly as the number of nodes is augmented; on the other hand, it turns out to be nearly constant around 0.85–0.95 ms. The delay for detection is nearly independent of the channel under use. It diverges somewhat as messy communication is used, which changes the detection BER with changes in the channel type.

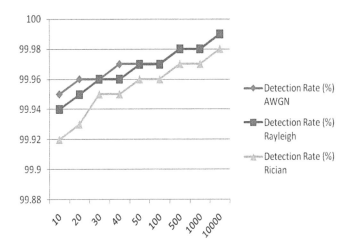

FIGURE 10.3
Recognition rate (%) vs. number of nodes.

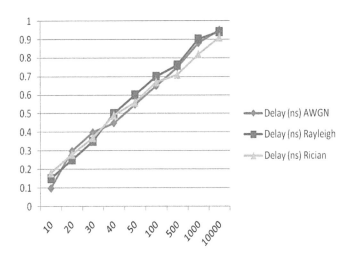

FIGURE 10.4
Delay in recognition vs. number of nodes.

10.3 Conclusions

In this chapter, an algorithm for primary user emulation attack detection and removal in cognitive radio networks is discussed. The approach makes use of the look-up table-based challenge sequences, which are monitored by the cognitive base station and act as the first line of defense against any primary emulation attacker. This ensures that almost all of the attackers are suppressed, and for the remaining attackers if any, a tag-based chaotic communication system is used, wherein each of the requests from secondary users are sent like a chaotic noise sequence on the channel, and the receiving entity decodes this sequence in order to get the signal communicated by an authorized transmitter.

If there is any communication by an attacker, then it is detected immediately, as none of the receiving entities can decode the signals sent by these unwanted attacker nodes. This ensures that the system guarantees greater than 99% detection and identification of attackers in primary user emulation attacks. The overall system accuracy is about 99% under different channel scenarios and varying node numbers. As the delay of detection is very less, this network can be used in real-time cognitive radio environments and will be helpful for hardware implementation of the two-layered protocols. More attacks can be added to the system to check the performance of the two-layer model under those attacks.

Furthermore, additional attack removal strategies for various other attacks can be realized because the implementation can detect the attackers in a very short span of time. Moreover, FPGA or IoT level implementation of the algorithms can be done to test the performance of the algorithms in real time under practical network conditions.

References

1. Y. Liu, P. Ning, H. Dai. 2010, Authenticating primary users' signals in cognitive radio networks via integrated cryptographic and wireless link signatures. *2010 IEEE Symposium on Security and Privacy*, Oakland, CA, pp. 286–301.
2. X. Tan, K. Borle, W. Du, B. Chen. 2011. Cryptographic link signatures for spectrum usage authentication in cognitive radio. In *WISEC'11: Fourth ACM Conference on Wireless Network Security*, Hamburg Germany, Association for Computing Machinery, New York, pp. 79–90.
3. M. Haghighat, S.M.S. Sadough. 2012. Cooperative spectrum sensing in cognitive radio networks under primary user emulation attacks. *Telecommunications (IST), 2012 Sixth International Symposium on*, Tehran, Iran, pp. 148–151.
4. M. Haghighat, S.M.S. Sadough. 2014. Cooperative spectrum sensing for cognitive radio networks in the presence of smart malicious users. *AEU-International Journal of Electronics and Communications*, 68(6): 520–527.
5. R. Chen, J.M. Park, J.H. Reed. 2008. Defense against primary user emulation attacks in cognitive radio networks. In *IEEE Journal on Selected Areas in Communications*, 26(1): 25–37.
6. Y. Zheng, Y. Chen, C. Xing, J. Chen, T. Zheng. 2016. A scheme against primary user emulation attack based on improved energy detection. *2016 IEEE International Conference on Information and Automation (ICIA)*, Ningbo, China, pp. 2056–2060.
7. H.H. Kuo, T.L. Liao, J.S. Lin, J.J. Yan. 2009. A new structure of chaotic secure communication in wireless AWGN channel. *Chaos-Fractals Theories and Applications, 2009. IWCFTA '09. International Workshop on*, Shenyang, China, pp. 182–185.

8. N. Nguyen-Thanh, P. Ciblat, A.T. Pham, V.T. Nguyen. 2015. Surveillance strategies against primary user emulation attack in cognitive radio networks. In *IEEE Transactions on Wireless Communications*, 14(9): 4981–4993.

9. J. Gu, S.H. Sohn. J.M. Kim, M. Jin. 2009. Chaotic characteristic based sensing for cognitive radio. *2009 5th International Conference on Wireless Communications, Networking and Mobile Computing*, Beijing, China, pp. 1–4.

10. A.A. Sharifi, M. Sharifi, and M.J.M. Niya. 2016. Secure cooperative spectrum sensing under primary user emulation attack in cognitive radio networks: Attack-aware threshold selection approach. *AEU-International Journal of Electronics and Communications*, 7(2–4): 95–104.

11. R. Zhou, X. Li, J. Zhang, Z. Wu. 2011. Software defined radio based frequency domain chaotic cognitive radio. *2011 IEEE International SOC Conference*, Taipei, Taiwan, pp. 259–264.

12. L. Zhang, J. Yu, Z. Wu. 2015. Secured chaotic cognitive radio system using advanced encryption standard. *2015 IEEE 26th Annual International Symposium on Personal, Indoor, and Mobile Radio Communications (PIMRC)*, Hong Kong, pp. 7–11.

13. Z. Jin, S. Anand, and K. P. Subbalakshmi. 2009. Detecting primary user emulation attacks in dynamic spectrum access networks, *IEEE International Conference on Communications*, IEEE, Dresden, Germany, pp. 1–5.

14. A. Riaz, M. Ali. 2008. Chaotic communications, their applications and advantages over traditional methods of communication, *Communication Systems, Networks and Digital Signal Processing, 2008. CNSDSP 2008. 6th International Symposium on*, Graz, Austria, pp. 21–24.

15. Y. Chen, L. Yang, S. Ma and X. Yuan. 2016. Detecting primary user emulation attacks based on PDF-BP algorithm in cognitive radio networks, *2016 IEEE International Conference on Internet of Things (iThings) and IEEE Green Computing and Communications (GreenCom) and IEEE Cyber, Physical and Social Computing (CPSCom) and IEEE Smart Data (SmartData)*, Chengdu, China, pp. 660–666.

16. M. Ghaznavi, A. Jamshidi. 2017. Defence against primary user emulation attack using statistical properties of the cognitive radio received power. *IET Communications*, 11(9): 1535–1542.

17. Y. Li, C. Han, M. Wang, H. Chen, L. Xie. 2016. A primary user emulation attack detection scheme in cognitive radio network with mobile secondary user. *2016 2nd IEEE International Conference on Computer and Communications (ICCC)*, Chengdu, China, pp. 1076–1081.

18. L. Liu, Z. Zhang, J. Li. 2016. EM-based algorithm for defeating against primary user emulation attacks in cognitive wireless networks. *2016 2nd IEEE International Conference on Computer and Communications (ICCC)*, Chengdu, China, pp. 1450–1455.

19. S. Madbushi, R. Raut, M.S.S. Rukmini. 2018. Trust establishment in chaotic cognitive environment to improve attack detection accuracy under primary user emulation. *Iranian Journal of Science and Technology, Transactions of Electrical Engineering*, 42: 291.

11

Cognitive Radio Based i-Voting System: An Application of Cognitive Radio Network

11.1 Introduction

In a traditional or conventional, paper-based election, the electorates cast their votes to select their representatives, where they simply deposit their designated ballots in sealed boxes distributed across the electoral circuits around a given country. By the end of the election period, all these boxes are officially opened and votes counted manually in the presence of certified representatives of all the candidates until the numbers are compiled. This process warrants transparency at vote casting time as well as at counting time [1]. Very often, however, counting errors take place, and in some cases, voters find ways to vote more than once, introducing irregularities in the final count results, which could, in rare cases, require a repeat of counting process or the election process altogether. Moreover, in some countries, purposely introduced manipulations of the electoral votes take place to distort the results of an election in favor of certain candidates. Here, all such mishaps can be avoided with a properly scrutinized election process; however, when the electoral votes are too large, errors can still occur.

Quite often, international monitoring bodies are required to monitor elections in certain countries. This naturally calls for a fully automated online computerized election process [2]. In addition to overcoming commonly encountered election pitfalls, electoral vote counts are done in real time such that by the end of elections day, the results are automatically out. The election process can be easily enhanced with various features based on the demand and requirements of different countries around the world. Due to worldwide advancements in computer and telecommunication technologies and the underlying infrastructures, online voting or i-Voting is no longer an American or Western phenomenon. The introduction of electronic voting has been the biggest change to the Irish electoral system since the establishment of the state over 80 years ago [3]. e-Voting may soon become a global reality or a global nightmare. The significant initiatives to examine the feasibility of electronic voting (including electronic voter registration) recently took place in the United Kingdom (www.peopletracer.co.uk). The United States has already used Internet voting program, that is, Federal Voting Assistance Program (FVAP) for their presidential election in 2016. FVAP works to ensure that the service members, their eligible family members, and overseas citizens are aware of their right to vote and have the tools and resources to do it successfully from anywhere in the world (www.fvap.gov).

In this chapter, cognitive radio (CR) based i-voting system is discussed. CR is the term first coined by J. Mitola et al. in the year 1999 for efficient spectrum utilization [4]. These CR devices are capable to change their operating parameters on-the-fly and adapt with respect to the environment. By doing so, CR devices opportunistically exploit the spectrum left unoccupied by primary radio (PR) nodes. Moreover, CR devices operate in an ad-hoc fashion and form CR ad-hoc networks. In this chapter, a framework that exploits CR technology is explored. This framework provides robust connectivity and Internet access to partially destroyed networks. In this context, to allow CR devices to restore the connectivity of partially destroyed coexistent network and at the same time to provide Internet accessibility, an architectural frame work is required, which in turn provides rapid, cost-effective, and robust connectivity.

The most crucial factor for a system like i-voting system to be successful is to exhibit a voting protocol that can prevent opportunities for fraud or for sacrificing the voters policy along with the reduction in money spent on election. CR-based i-voting is an Internet-based online voting system. It works same as that of other online services like online reservation system. In this, people who are Indian citizens and having age 18 years and above can cast their vote online without going to polling booth. In i-voting system, a voter can use his/her voting rights without any difficulty. However, the necessary thing to the voter is that he/she must have valid voter ID card. Aadhaar is a number that serves as a proof of identity and address, anywhere in India. Any individual, irrespective of age and gender, who is a resident in India and satisfies the verification process laid down by the Unique Identification Authority of India (UIDAI), can enroll for Aadhaar (www.uidai.gov.in).

11.2 CR: Touching the Inaccessible Ends

In rural area where Internet facility is not available to the people so easily, Cognitive Radio Ad-Hoc Networks (CRAHNs) is a promising technology and capable to provide the communication of coexistent networks temporarily [5]. In fact, several distinguished features of CR technology make CRN easy to deploy and flexible solution for challenged environments. These features include for instance the accessibility and flexibility of communication over the whole spectrum band. Another important feature is the multiradio capability of CRNs, which can further be used to control communication overhead. Despite the availability of whole communication spectrum band, concentration of all the radio devices over a single spectrum band could lead to contention and collision problems, which further reduces the connectivity to global Internet. CR could help a lot on these, by providing more "communication space" to devices. In addition, the inherent self-organizing capabilities of CR devices are to detect and choose among channels and features provided by the spectrum sensing and allocation.

Another CR-based network architecture DIMSUM net by Buddhikot et al. is proposed for the integration of CR networks into the global Internet [6]. This architecture is specifically designed for cellular networks and based on a centralized regional spectrum broker and does not cater the needs of rapid and ad-hoc network deployments, mostly required in Internet-based voting system. In addition, CR ad-hoc networks have been widely used in several application scenarios including military and mission critical networks, and consumer-based applications [7,8]. CR technology can also play an important role in E-health applications [9].

11.3 Need of Voting Process Authentication and Voter's Right Privacy

Certain factors play out big role in a given voting process, particularly in a country like India. Culture itself and the underpinning social factors/ values largely determine the rules and regulations that govern any voting process. In countries, where election results are determined through the voter counts that are tallied by directly depositing specially designed voting cards into the voting boxes, there are tendencies that electoral votes can get disqualified in many ways; some voters would tend to attempt to vote more than the number of times permissible by law for a given candidate; other voters may try to vote in lieu of other illegible voters so that the voter count would weigh favorably toward one candidate or another, to mention just a few. Counterfeit/malice is yet another issue that can jeopardize the integrity of an election process. Automating an election process, while relying on state-of-the-art in computer technologies, can significantly mitigate many of the factors that would hamper a healthy progress of the election process. Nonetheless, relying totally on available information technologies can only warrant the authentication/validation of the identity of a given voter but still would not have the capacity to block any attempted malpractices of the voting system, viz., those voters who simply try to vote on behalf of others (fraud) [10]. Without additional measures, the integrity of a voting process, within the proper context, is far from any acceptable standard/s; the incorporation of biometrics would definitely have an added value toward achieving the required levels of election integrity. Present day applications, including banking applications, guarding of high-security establishments, monitoring of passengers across border posts, among many others are witnessing increasing levels in the use of biometric technologies and devices. Biometrics is best defined as measurable physiological and/or biological characteristics that can be utilized to verify the identity of an individual [11]. They include fingerprints, retinal and iris scan, hand geometry, voice patterns, facial recognition, Gait recognition, DNA, and other techniques. They are of interest in any area where it is important to verify the true identity of an individual.

Initially, these techniques were employed primarily in specialist high-security applications; however, they are now used in a much broader range of public facing situations. Essentially, a biometric system follows two characteristic traits: identification and verification. The former involves identifying a person from all biometric measurements collected in a database. The question that this process seeks to answer is: "who is this?" It, therefore, involves a one-compared-to-many match. Verification involves authenticating a person's claimed identity from his/her previously enrolled pattern. "Is this who he claims to be?" is the question that this process seeks to answer. This involves a one-to-one match. Verifying the identity of a person against a given biometric measure involves five phases that the system needs to go through.

In i-voting, 12-digit Aadhaar ID number and facial expression as a biometric measure are used because UIDAI has already stored captured photographs in their database for authentication and verification purpose. At the beginning, input data is read from the person through the reading sensors. Collected data is then sent across a network to some central database hosting a biometric system. The system will, then, perform identity matching using standardized and/or custom matching techniques. Figure 11.1 illustrates data flow in a typical biometric identification process.

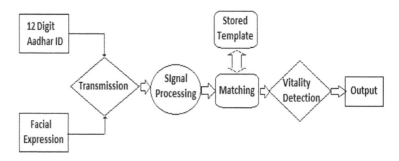

FIGURE 11.1
Biometric system data. (From R.V. Awathankar et al., *2016 IEEE Inter. Conf. Comput. Intell. Comp. Res.*, 1–6, 2016.)

11.4 The i-Voting System

The i-voting system is mainly divided in two parts. First part gives information about i-voting system and client–server architecture. The second part shows how this i-voting system can be implemented in rural part where Internet facility is not easily available using CRN technology.

In this i-voting system, people can cast their vote through the Internet. In order to prevent voter frauds, two-level voter authentication is used for security purpose. Aadhaar ID is used as a first level of authentication where a voter needs to enter his/her 12-digit Aadhaar ID number. The Aadhaar ID entered by the voter is verified with the contents of the E-Aadhaar server database. If valid ID is entered, then the voter will be forwarded to second level of authentication process. In second level authentication, face recognition process will carried out. Here, the facial expression of the voter will be captured by a web camera and sent to the E-Aadhar database at server side for verification.

After the successful authentication and verification, the voter should cast his\her vote now by selecting his/her registered region. The web page could be designed using any of the available resources like .NET, JAVA, PHP, etc. This web page is connected directly to Aadhaar server database. Figure 11.2 shows the entire process of i-voting system data flow.

Client/server web-enabled i-voting software architecture is illustrated in Figures 11.3 and 11.4. Besides the main functional properties of a voting system, as described in the previous section, the i-voting system must cater for several essential nonfunctional requirements. Of utmost importance are the requirements for correctness, robustness, coherence, consistency, and security. On the server side, an Aadhaar database is maintained by UIDAI for all registered voters and candidates. Also, the server runs in real-time and provides backend statistics for the entire election process.

On one side, the system prints a hardcopy of the vote cast by the voter. The voter verifies the accuracy of his/her vote and retains the copy for his/her records. On the other hand, the system generates another copy of the vote with a new unique key identifier; the name and identity of the voter is concealed. This copy is saved in a secure box and can be used later to verify the correctness of the votes as stored in the final DB destination. This side of the copy can be printed out as a bar code, which can be easily scanned and read automatically. Only a randomly selected set of these copies need to be tested. This two-sided process guarantees transparency by providing verification of the accuracy of how the cast vote is input into the system and then how it is, finally, stored in the DB tables.

Internet access framework for future CRNs is a three-tier architectural framework tailored to implement and deploy real CRN applications in rural communication environments. Figure 11.5 depicts the concept of

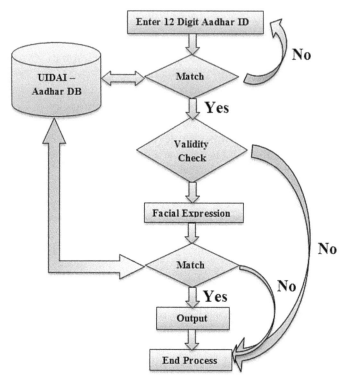

FIGURE 11.2
i-voting system data flow. (From R.V. Awathankar et al., *2016 IEEE Inter. Conf. Computa. Intell. Comp. Res.*, 1–6, 2016.)

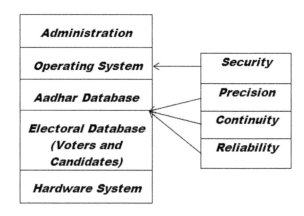

FIGURE 11.3
(Server) (From R.V. Awathankar et al., *2016 IEEE Inter. Conf. Computa. Intell. Comp. Res.*, 1–6, 2016.)

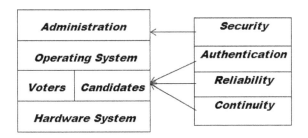

FIGURE 11.4
(Client) (From R.V. Awathankar et al., *2016 IEEE Inter. Conf. Computa. Intell. Comp. Res.*, 1–6, 2016.)

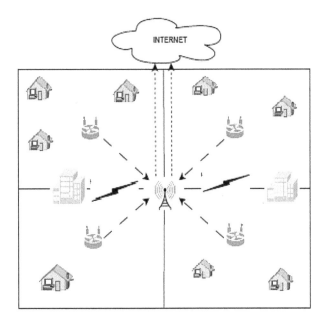

FIGURE 11.5
Internet Access Framework for CR-based i-voting system. (From R.V. Awathankar et al., *2016 IEEE Inter. Conf. Computa. Intell. Comp. Res.*, 1–6, 2016.)

internet access framework for CR-based i-voting system. The building blocks of this architecture are:

1. Cognitive multiradio mesh routers (CMR)
2. Internet portal point
3. Public polling booth

In this architecture, the Internet portal point is the primary network and their nodes are primary nodes. Indeed, the primary objective is to offer them connectivity to the Internet. It is clear that interconnecting different type of networks using different technologies can be considered as a challenging task; however, the flexibility and dynamic spectrum management offered by CRN can help to overcome these obstacles. Internet portal points are connected directly to the satellite as a secondary user. The polling is held normally from 7:00 am to 5:00 pm. For such small period of time, it will be very costly to buy spectrum. Instead, the Electoral Commission should have a tie up with particular band from any fixed channel so that the system should have access to Internet as a secondary user during the electoral period.

Internet portal points are devices that serve as gateways to the Internet. These devices can be stationary or mobile; equipped with powerful communication medium, for example, satellite link. These are responsible for sharing Internet bandwidth as well as gathering data from cognitive multiradio mesh routers and transfer it to the Internet. Initially, the network is deployed having a single Internet portal point device. This device should be connected with the global Internet through the satellite link. In the vicinity of this Internet portal point, fixed cognitive multiradio mesh routers are deployed, which are directly connected with the Internet portal point. Internet portal point shares the Internet connection with these cognitive multiradio mesh routers. The voters have non-CR devices like PC and laptop and can connect to the Internet through multiradio mesh routers. However, it is not possible to have non-CR devices with every user. So, to avoid this problem, public polling booth, which is optional, should also be made available at every particular parts of the region.

 Cognitive Multi-Radio Mesh Router

 Internet Portal Point

 House

Public Polling Booth (Optional)

11.5 Conclusions

In present scenario, government spends a huge amount for conducting elections. This money is spent on issues such as security, electoral ballots, etc. The average percentage of voting is a less than 60%. Moreover, voting fraud can be easily done in the present system. Also, the percentage of literates coming to vote is very less. There is a strong need to curtail the money spent on election, and at the same time it is necessary to control voting fraud. Moreover, security is also of paramount importance.

In this chapter, a CR technology is presented as a potential candidate to curb all the menaces discussed above. In this regard, a CR-based Internet access framework for rural area is envisaged using the Aadhaar card. This architecture specially caters to the needs of challenged environments. Furthermore, the issues and challenges in the deployment of this architecture are highlighted.

References

1. M.G. Margaret, M.C. Joe. 2004. Transparency and eVoting: Democratic vs. commercial interests. www.cs.nuim.ie/~mmcgaley/Download/Transparency.pdf.
2. R. Mercuri. 2000. Electronic vote tabulation checks and balances. PhD thesis, University of Pennsylvania, Philadelphia, PA.
3. M.G. Margaret. Irish Citizens for Trustworthy Voting. 6 July 2004. http://evoting.cs.may.ie/.
4. J. Mitola, G.Q. Maguire. 1999. Cognitive radio: Making software radios more personal. *IEEE Personal Communications*, 6(4): 13–18.
5. I.F. Akyildiz, W.Y. Lee, K.R. Chowdhury. 2009. CRAHNs: Cognitive radio ad hoc networks. *Ad Hoc Networks*, 7(5): 810–836.
6. M.M. Buddhikot, P. Kolodzy, S. Miller, K. Ryan, J. Evans. 2005. Dimsumnet: New directions in wireless networking using coordinated dynamic spectrum access. In *Sixth IEEE International Symposium on World of Wireless Mobile and Multime- dia Network (WoWMoM)*, pp. 78–85.
7. S. Ball, A. Ferguson, T.W. Rondeau. 2005. Consumer applications of cognitive radio defined networks. In *First IEEE International Symposium on New Frontiers in Dynamic Spectrum Access Networks (DySPAN' 05)*, pp. 518–525.
8. O. Younis, L. Kant, K. Chang, K. Young. 2009. Cognitive manet design for mission-critical networks. *IEEE Communications Magazine*, pp. 64–71.
9. S. Feng, Z. Liang, D. Zhao. 2010. Providing telemedicine services in an infrastructure-based cognitive radio network. *IEEE Wireless Communications Magazine*, pp. 96–103.
10. A.D. Rubin. 2002. Security considerations for remote electronic voting. *Communications of the ACM*, 45(12): 39–44. http://avirubin.com/e-voting.security.html.
11. S. Nanavati, M. Thieme, R. Nanavati. 2002. *Biometrics: Identity Verification in a Networked World*. John Wiley & Sons, New York.

12

Sparse Code Multiple Access (SCMA) for Cognitive Radio: An Introduction

12.1 Introduction

The requirement to connect large number of users efficiently led to the refinement in wireless communication from second generation to fifth generation. The enhancement of cellular communication system involves use of various multiple accesses.

5G wireless communication aims at improving the spectral efficiency for connecting massive users. In order to satisfy the above demand new spectrum bands are required. Due to scarcity in the spectrum resources deployment of cognitive radio features in 5G networks can provide avoidance in the interference with other operating networks.

To cater this problem, orthogonal frequency division multiple access (OFDMA) is used with cognitive radio to dynamically avoid interference. Another approach is MC-CDMA, multicarrier-code division multiple access, which deals with the orthogonality of spreading sequences. Using cognitive radio in 5G wireless communication results into study of one of the nonorthogonal multiple access (NOMA) technique, that is sparse code multiple access (SCMA). The cognitive features are

- Spectrum sensing
- Interference estimation
- Subcarrier adaptability

These can easily be improved in SCMA for 5G communication.

This chapter aims at elaborating the multiple access techniques progressing from 2G to 5G and describes SCMA one of the NOMA techniques used in 5G cellular communication system [1]. It is a codebook-based multidimensional nonorthogonal spreading technique. It involves combining of quadrature amplitude modulation (QAM) mapper and spreading signature blocks into a single block of a SCMA spreading encoder having SCMA codebook set that results in a multidimensional codeword.

The 5G wireless communication involves diverse applications, which will be deployed by 2020. The most important requirement of 5G is its high spectral efficiency. Apart from that, high throughput, better service, quality, low control signaling, and lower latency are some of the requirements to be met while using any access. In a cellular system, the channel bandwidth is limited, whereas it has to accommodate maximum users in it; thus, multiple access is a technique that helps the cellular communication to be more economical by maximum utilization of channel bandwidth as a physical layer technology. It enables the wireless base stations to identify various users and serve them.

The different ways that allowed access to the channel included mainly orthogonal and nonorthogonal access. In orthogonal access, the cross correlation of signals from different users is zero, which can be achieved by frequency division multiple access (FDMA), time division multiple access (TDMA), code division multiple access (CDMA), and orthogonal frequency division multiple access (OFDMA) nonorthogonal schemes allow nonzero cross correlation among the signals from different users, such as in random waveform code-division multiple access (CDMA), Trellis-coded multiple access (TCMA), and interleave division multiple access (IDMA). Power domain multiple access, low-density signature OFDM (LDMA), pattern division multiple access (PDMA), building block sparse constellation based multiple access (BOMA), sparse code multiple access (SCMA), lattice partition multiple access (LPMA), and many more. The multiple accesses used in communication from 2G to 5G are shown in Figure 12.1.

The 2G communication system made use of basic multiple access [2], that is, TDMA and FDMA wherein the users are scheduled on orthogonal time slots. TDMA is a multiple access method that allows different users to use the same channel bandwidth by dividing the transmitted signals from the users into the different time slots.

In TDMA, channel bandwidth is orthogonal to time in time–frequency and code domain.

FDMA is a multiple access method in which the channel bandwidth is divided completely according to the number of users. Thus, the complete channel bandwidth is utilized by the user for the specific time period.

In FDMA, the channel bandwidth is orthogonal to frequency in time–frequency and code domain.

The 3G communication system later made use of CDMA where the channels are nonorthogonal in frequency and time domain but orthogonal in code domain. CDMA is a multiple access method where different transmitters can send the signals simultaneously over the same channel bandwidth.

In CDMA, the codes are orthogonal to frequency and time in time–frequency and code domain.

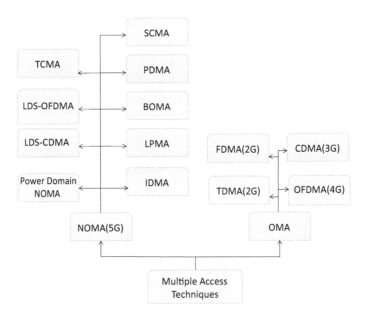

FIGURE 12.1
Multiple access techniques used in the communication system.

The 4G communication systems, widely known as long-term evolution (LTE), make use of OFDMA, where the users are orthogonal in 2D frequency–time domain. Orthogonal resources are occupied by the users for communication. Being the access with single user detection, it becomes comparatively easy to implement. Here, the subsets of subcarriers are assigned to the individual user. It is derived from OFDM, that is orthogonal frequency division multiplexing, that aims at allocating the users in time domain only, whereas OFDMA aims at allocating the users in both time and frequency domains.

As a result of limited users of orthogonal resources, which is proportional to the number of users, 5G communication system led to the extensive research on various multiple accesses, that can meet the three basic demands as per 3G Partnership Project (3GPP), which includes massive machine type communication (mMTC), ultra reliable and low latency communication, and enhanced vehicle-to-everything (eV2X) communications.

To achieve the same, 5G communication systems require large connectivity with good throughput and spectral efficiency. These challenges can be addressed by introducing NOMA techniques.

Table 12.1 shows the comparison of frequently used schemes in NOMA with their multiplexing advantages and disadvantages. In a communication system various channel properties of the communication link are referred to as channel state information (CSI), which is one of the important parameters responsible for the quality of wireless communication systems.

TABLE 12.1

Comparison of Various Access Techniques in NOMA

Schemes	Characteristics	Advantages	Disadvantages
Power-Domain NOMA	Power domain multiplexing	High SE, Compatible to other techniques	Need user pairing error propagation in SIC
LDS-CDMA	Sparse spreading CDMA	No need of CSI, near optimal MPA detector	Redundancy from coding
SCMA	Sparse spreading multidimensional constellation	No need of CSI, near optimal MPA detector, more diversity than simple LDS	Redundancy from coding, difficult to design optimal code book
LDS-OFDM	Sparse spreading OFDM	No need of CSI, near optimal MPA detector, more fit for wide-band than LDS-CDMA	Redundancy from coding
PDMA	Sparse spreading multiplexing in power code and spatial domain	More diversity, near optimal MPA detector, low complexity receiver	Redundancy from coding, difficult to design optimal patterns
BOMA	Tiled building block	Simple structure compatible to current system low complexity receiver	Need user pairing, not very flexible
LPMA	Multilevel lattice code, multiplexing in power and code domains	No need of user clustering	Specific channel coding

If CSI is not estimated properly, cross-layer interference limits the potential performance gain of MU-MIMO.

If we compare various NOMA techniques listed in Table 12.1, we can conclude that LDS-CDMA, LDS-OFDMA, and SCMA are the techniques where CSI is of least importance, hence more desirable. In this chapter, we focus more on the study of SCMA.

SCMA does not require the CSI for the transmitter and the receiver in the communication link. It is also responsible for reducing the interference in MU-MIMO to enhance the link performance. It enables grant-free transmission with low overhead and low latency for sporadic small packet transmission. The QAM modulation and the LDS spreading are replaced by multidimensional codebooks in SCMA. Despite large number of users, the collisions are less and have better coverage because of the spreading gain. Table 12.2 gives the comparison of main aspects of LDS and SCMA, which makes use of message passing algorithm (MPA) detector at the receiver with same structure and complexity over the codewords and symbols [3].

TABLE 12.2

SCMA vs LDS

Schemes	SCMA		LDS	
Multiple access	Yes	Codebook domain	Yes	Signature domain
Sparse	Yes	Sparse codewords	Yes	Low-density signatures
Coding gain	Yes	Data carried over multi-dimensional complex codewords	No	Data carried over QAM symbols
Degree of freedom	J codebooks each with M codewords		J signatures	
Receiver	Codeword-based MPA		Symbol-based MPA	

12.2 Sparse Code Multiple Access

SCMA is a multidimensional codebook-based access, based on nonorthogonal technique [4,5], introduced by Huawei Technologies. In this access, the incoming bits are directly mapped into multidimensional codewords and are transmitted across the channel. The implementation of link-level simulation for SCMA requires only few modifications to be done on the LTE transceiver. At the receiver end to reduce the complexity of decoding and avoid the interference of the channels, MPA is used.

SCMA is a technique used in 5G communication system for economic, energy-efficient link layer performance and low complexity implementation. CDMA is a multiple access in which the data is spread out over orthogonal code sequences. Low-density signature (LDS) is a more advanced approach of CDMA. In CDMA, the CDMA signature expands to a QAM symbol generated from QAM mapper. However, SCMA involves clubbing of QAM mapper and CDMA spreader as shown in Figure 12.2.

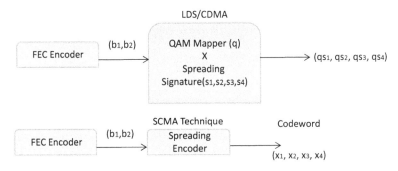

FIGURE 12.2
Clubbing of QAM mapper with CDMA spreader.

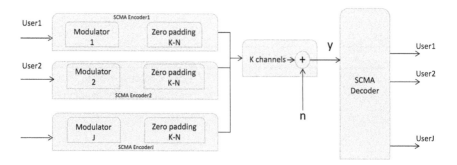

FIGURE 12.3
Fixed uplink SCMA system model.

At transmitter, the multidimensional codewords are formed by mapping the coded bits directly in complex domain, and codewords from different users are overlapped nonorthogonally in sparse spreading way. Signal detection is done by the receiver, which is then followed by channel decoding for data recovery (Figure 12.3).

12.3 SCMA System Model

There are J user layers in the system. The signal of each user layer is spread into k orthogonal resource layers. SCMA transmits the signal in an overloaded manner, that is, λ. The log $|M|$ input binary bits b_j with unique codebook $X_j \subset \mathbb{C}k$ are mapped into a k-dimensional codewords X_j. The codewords mapped by input bits correspond to the codebook of the user layer where different codebook is owned by different users. The design of encoder involves two steps as given below:

1. Binary bits b_j are modulated to d bits from ($d0$: dN)
2. where g_j will be the modulating generator matrix represented by

$$d_j = b_j\left(g_j\right) \tag{12.1}$$

3. K-N zeros are mapped in the X_j codeword, which is k-dimensional
4. with $V_j \in B$, $k*N$ matrix, thus expressed as

$$X_j = V_j\left(d_j\right). \tag{12.2}$$

The design of tanner graph and the sparse matrix is as shown in Figure 12.4.

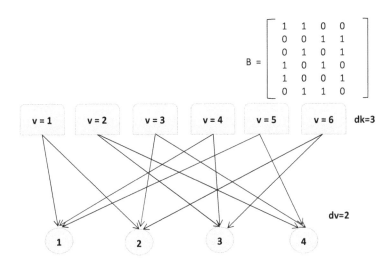

$$B = \begin{bmatrix} 1 & 1 & 0 & 0 \\ 0 & 0 & 1 & 1 \\ 0 & 1 & 0 & 1 \\ 1 & 0 & 1 & 0 \\ 1 & 0 & 0 & 1 \\ 0 & 1 & 1 & 0 \end{bmatrix}$$

FIGURE 12.4
Tanner graph and matrix.

TABLE 12.3

Variables used in Tanner Graph

V	6	No. of layers
F = k	4	Number of functional nodes, i.e., available subcarriers
M	4	Length of codeword in number of bits
Dv	2	Maximum k nodes connected to each V node (No. of used subcarriers)
Dk	3	Maximum V nodes connected to each k node
Λ	(k−1)/2 = 1.5	Overloading factor

Number of 1s in each column is denoted by dk, which is the number of users from which each subcarrier accepts the data, whereas the number of 1s in each row is denoted by dv. Table 12.3 shows the variables used in the tanner graph.

12.4 Design of Codebook

The design of codebook for SCMA is supposed to be joint optimization of multidimensional constellation design, which makes this access more unique than the other NOMA techniques. The basic aim of the codebook design is to maximize the shaping gain by providing good distance properties in the multidimensional constellation.

It should also have less projection points over each resource element. One can also design a multidimensional codebook on the basis of 8-QPSK constellation. Another approach of designing an efficient codebook is using star-QAM signaling constellation, which helps in improving the BER without hampering low detection complexity. A multidimensional SCMA codebook design can be on constellation rotation and interleaving, which proposes lower BER than the LDS in down-link Rayleigh fading channels. The spherical code method is another approach that improves the system performance in order to build mother multidimensional codebook. The designing equation of the codebook for SCMA is given by

$$V^*, G^* = \arg \max V, Gm\big(S(V, G; J, M, N, K)\big) \qquad (12.3)$$

where m is the design criterion; in order to achieve suboptimal solution for this multidimensional problem, a multistage optimization approach is proposed.

SCMA aims at reducing the complexity of SCMA detectors. The detection can be realized by finding the maximum joint posterior probability of all users' transmitted symbols. As a result of enormous computations, practical implementation is hampered. Hence, the MPA, which is used in LDS based on the sum–product algorithm, is used by SCMA detector to reduce computation.

In order to reduce the complexity, fixed-point implementation of the log-domain message passing algorithm (Log-MPA) was implemented. Another proposed approach used two receivers, which not only simplified the detection structure but also curtailed exponent operations quantitatively in logarithm domain. Low complexity of detection and low complexity detection structure due to the codebooks is studied for the receiver. Fixing t codewords in the mth iteration is a simplified detector based on partial marginalization (IPM-MPA) detector for the fixed uplink SCMA system. Thus, IPM-MPA is more computational efficient than PM-MPA.

12.5 SCMA Link Implementation

By doing minor modifications in the LTE transceiver, the link-level simulations for SCMA are implemented. As shown in Figure 12.5, the change made in the transmitter is done by replacing the QAM modulator and the DFT block by SCMA encoder, which maps the coded bits into multidimensional codeword. The spreading factor of these transmitted bits can be seen from the tanner graph. As seen in the transmitter only by making few modifications in the LTE receiver, the receiver of SCMA is designed by replacing single-user channel equalization and QAM demapper with SCMA decoder of each layer. Tanner graph constructed by the codebooks are performed by the MPA. It starts with the initial conditional probability calculation at each function node.

Then, it enters MP iterations between the function node and the variable node. For each iteration, both the nodes are updated; this is done

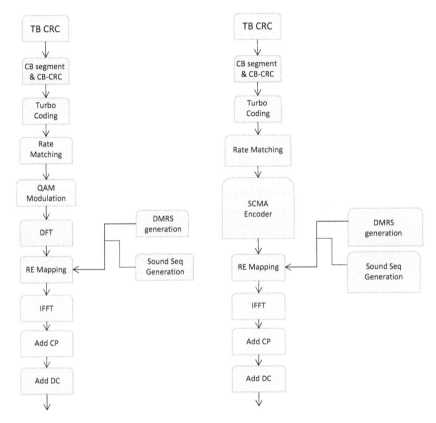

FIGURE 12.5
Uplink implementation for SCMA at the transmitter end as compared to OFDMA transmitter.

independently by each pair. After several sufficient iterations LLR (Log Likelihood ratio) for coded bits are calculated based on codeword probability and output at variable node and can be given as the input to the turbo decoder. Thus, the user-to-user communication in SCMA can be studied.

12.6 Codebook Design Methods

The requirements of 5G communications include more substantial connectivity with low latency, including massive data traffic with reduced bit error rate. OFDMA is a technique used by 4G wireless communication, which fails to serve these future requirements of 5G communications, thus leading to the research of various NOMA techniques. As per the comparative study of different NOMA techniques, SCMA proves to be the most reliable candidate in NOMA techniques for 5G communications. SCMA directly maps the incoming bits to complex multidimensional codeword selected from predefined codebook sets. This technique is designed from the LDS-based CDMA approach (LDS-CDMA) but has better spectral efficiency and better shaping gain or coding gain of multidimensional constellation. In this technique, the QAM mapper and the CDMA spreader are clubbed together, and a set of bits are mapped directly into a complex codeword, selected from predefined codebook set. A specific codebook set has been defined for every layer. These layers are placed nonorthogonally above each other.

At the receiver end, the MPA detector is used with moderate complexity. The sparseness of codewords enables the MPA to alleviate the interuser interference with reduced complexity. The BER, bit error rate of SCMA, is primarily dependent upon the codebooks, but designing a codebook is comparatively a more significant problem still under study.

The fundamental aim of the codebook design is to maximize the shaping gain by providing good distance properties in the multidimensional constellation. Following steps are involved in designing a codebook [4].

- To generate multidimensional modulation constellations
- To transfer the constellation to multiple sparse codebooks

12.7 Conclusions

This chapter explains the progress of wireless communication toward SCMA as a promising technology for 5G wireless communication and cognitive radio technology in particular.

References

1. Y. Chen, A. Bayesteh, Y. Wu, M. Taherzadeh, D. Chen, J. Ma, S. Han. 2016. SCMA: A promising non-orthogonal multiple access technology for 5G networks. *2016 IEEE 4th Vehicular Technology Conference (VTC-Fall)*.
2. P. Wang, J. Xiao, L. Ping. 2006. Comparison of orthogonal and non-orthogonal approaches to future wireless cellular systems. *IEEE Vehicular Technology Magazine*, 1(3): 4–11.
3. K. Sunil, P. Jayaraj, K.P. Soman. 2012. Message passing algorithm: A tutorial review OSR. *Journal of Computer Engineering (IOSRJCE)*, 2(3): 12–24.
4. M. Taherzadeh, H. Nikopour, A. Bayesteh, H. Baligh. SCMA codebook design. *2014 IEEE 80th Vehicular Technology Conference* (VTC2014-Fall). IEEE, 2014.
5. H. Nikopour, H. Baligh. 2013. Sparse code multiple access. *IEEE 24th International Symposium on Personal, Indoor and Mobile Radio Communications: Fundamentals and PHY Track*.

Appendix A.1: GNU Radio Installation Procedure

1. Install Ubuntu 14.04: Either dual-boot, native install or in virtual machine.

2. Update OS:

    2.1 Command in terminal: sudo apt-get update

    2.2 Command in terminal: sudo apt-get upgrade

3. Install dependencies for GNURadio:

    sudo apt-get install libfontconfig1-dev libxrender-dev

    libpulse-dev swig g++ automake autoconf libtool python-dev libfftw3-dev

    libcppunit-dev libboost-all-dev libusb-dev libusb-1.0-0-dev fort77 libsdl1.2-dev

    python-wxgtk2.8 git-core libqt4-dev python-numpy ccache python-opengl

    libgsl0-dev python-cheetah python-lxmldoxygen qt4-default qt4-dev-tools

    libusb-1.0-0-dev libqwt5-qt4-dev

    libqwtplot3d-qt4-dev pyqt4-dev-tools python-qwt5-qt4 cmake git-core wget

    libxi-dev python-docutils gtk2-engines-pixbuf r-base-dev python-tk

    liborc-0.4-0 liborc-0.4-dev libasound2-dev python-gtk2 zeroc-ice libcomedi-dev

    liblog4cpp5-dev libitpp-dev python-matplotlib wireshark gfortran liblapack-dev

    libblas-dev libpolarssl-dev subversion

    libopenblas-dev pandoc libtecla1 libtecla1-dev python-scipy

4. Download latest version of GNURadio: http://gnuradio.org/redmine/projects/gnuradio/wiki

5. Copy downloaded GNURadio in Documents folder: mv ~/Downloads/gnuradio-3.7.7.1.tar.gz ~/Documents

6. Unzip by right-click on tar file.

7. cd ~/Documents/gnuradio-3.7.7.1

8. mkdir build && cd build

9. cmake -DCMAKE_INSTALL_PREFIX=~/gnuradio../

10. make -j4
11. make install
12. cd ~
13. gedit ~/.bashrc
14. Goto the bottom the file and copy-paste the following command:
    GRPREFIX=~/gnuradio

    if [-z `echo $PATH | grep "$GRPREFIX"`];

    then

    export PATH=/usr/lib/ccache:$PATH:~/bin:$GRPREFIX/bin

    export   LD_LOAD_LIBRARY=$LD_LOAD_LIBRARY:$GRPREFIX/
       lib

    export LD_LIBRARY_PATH=$LD_LIBRARY_PATH:$GRPREFIX/lib

    export PYTHONPATH=$PYTHONPATH:$GRPREFIX/lib/python 2.7/
       dist-packages

    export PKG_CONFIG_PATH=$PKG_CONFIG_PATH:$GRPREFIX/
       lib/pkgconfig

    fi
15. Close the terminal and open again
16. Command in terminal: sudo ldconfig
17. gnuradio-companion

Appendix A.2: GNU Radio Sample Programs From Amitec Laboratory Manual (www.amitec.co)

EXPERIMENT 1: Custom Project Implementation: Developing and Testing Building Blocks of Communication Systems

The proposed experiment gives the basic understanding of using GNU Radio Companion (GRC), which is a graphical tool for creating signal flow-graphs and generating flow-graph source code. The GNU Radio is a free software development toolkit that provides the signal processing runtime and processing blocks to implement software radios using Amitec SDR04 Platform. This particular experiment will help students to learn basic blocks that are required in communication systems and their implementation using GRC will be discussed.

1.1 EXPERIMENT DETAILS: STEPS TO FOLLOW

This section describes how to start working in GNU radio platform for the beginners. Students should use the following steps as a beginner which are generally used in most of the GRC experiments:

Step 1: Open a terminal window using keyboard inputs: **Ctrl+Alt+T**, or by going to Dash Home on top left side and typing "Terminal" in it.

Step 2: At the terminal prompt type: **gnuradio-companion**.

Step 3: An untitled GRC window similar to the one shown in Figure A2.1.1 will open. If this window does not appear, then close all the windows and open a new window.

Step 4: Save this flow-graph.

Step 5: Double click on the Options block. This block sets some general parameters for the flow-graph. Leave the ID as top_block. Type in a project title (such as Experiment 1) and author. Set Generate Options to **QT GUI**, **Run** to **Autostart**, and **Realtime Scheduling** to **Off**. Then close the properties window.

Step 6: Open the other block named Variable block present in the flow-graph. It is used to set the sample rate. Set this equal to 32000.

FIGURE A2.1.1
An untitled GRC window appearing in the beginning.

Step 7: Students can find list of available blocks on the right side of the window. By expanding any of the categories (click on triangle to the left) students can see the available blocks. Students are advised to explore each of the categories to familiarize with the available blocks.

Step 8: Open the **Waveform Generators** category and double click on the **Signal Source**. Note that a Signal Source block will now appear in the main window. Double clicking on the block will open the properties window as shown in Figure A2.1.2. In this window, students can define sampling rate, frequency and amplitude of the signal. Sample rate should be chosen in accordance with Nyquist theorem.

Step 9: In order to view the waveform output from the source, students will have to choose a sink. From **Instrumentation category, add QT GUI Time Sink** which is under QT category. Change the type to float.

Step 10: In order to connect these two blocks, click once on the "out" port of the Signal Source, and then once on the "in" port of the Time Sink. The following flow-graph shown in Figure A2.1.3 will display.

Step 11: Connecting source directly to sink may consumes whole of the CPU processing due to which system may hang. Therefore, a block called **Throttle** needs to be added after source. This block limits the data rate by which data passes between the blocks. Select this block from **Misc** category. Change the type to float.

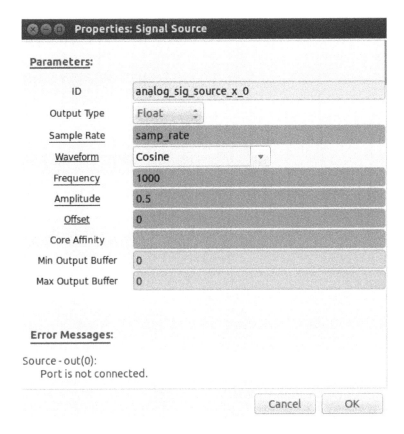

FIGURE A2.1.2
GUI window for setting signal source parameters.

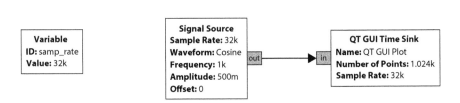

FIGURE A2.1.3
Flow-graph for displaying waveform output from the signal source.

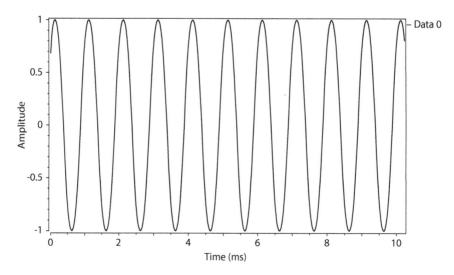

FIGURE A2.1.4
Plot displaying output signal captured from time sink.

Step 12: Click on "Execute the flow-graph" icon. A scope plot should open displaying several cycles of the sinusoid. Students should check the frequency and amplitude of this signal and it should match the value input in signal source properties in Figure A2.1.2. Such plot is displayed in Figure A2.1.4. One can play with the controls on the scope plot.

Step 13: Close the Scope Plot and change the sample rate back to 32000. Add a QT GUI Frequency Sink (under Instrumentation) to the window. Change the Type to Float and leave the remaining parameters at their default values.

Step 14: Connect this to the output of the Signal Source by clicking on the out port of the Signal Source and then the in port of the Frequency Sink. Generate and execute the flow graph. Student should observe the scope to verify the frequency of the input signal as 1 KHz. Close the output windows.

Step 15: Execute the flow-graph by pressing F6.

1.2 SAMPLE PROJECT FOR DEVELOPING FLOW-GRAPHS

Following two examples shows students how to develop their own flow graph, its execution and displaying various plots.

Experiment #1: Construct the flow-graph shown in Figure A2.1.5. Set the sample rate to 32000. The two Signal Sources should have frequencies of 1000 and 800 Hz, respectively. The block name **Add** can be found in the Operators category.

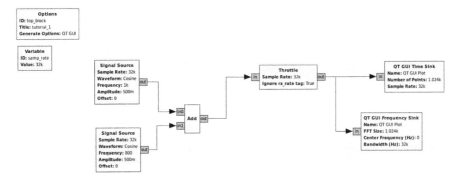

FIGURE A2.1.5
Example flow-graph for testing two tones signals.

Remarks: After generating and executing the flow-graph, the student can observe a waveform corresponding to the sum of two sinusoids on the scope plot. The time and frequency plots are shown in Figure A2.1.6. On the Frequency plot, students can see components at both 800 and 1000 Hz. Unfortunately, the Frequency plot does not provide enough resolution to clearly see the two distinct components. Note that the maximum frequency displayed on this plot is 16 KHz. This is one-half of the 32 KHz sample rate. In order to obtain better resolution, we can lower the sample rate. Try lowering the sample rate to 10 KHz.

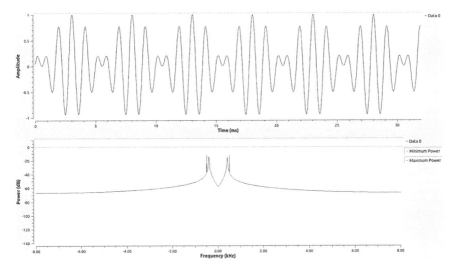

FIGURE A2.1.6
Time and frequency plot of example 1 flow-graph.

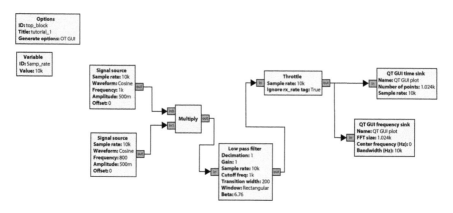

FIGURE A2.1.7
Flow-graph for example 2.

> *Student Exercise: Replace the ADD block with MULTIPLY block. What output one can expect from the product of two sinusoids? Confirm the result on the Scope and Frequency displays. Also change the sample rate to 10 KHz.*

Example 2: Modify the flow-graph of example 1 to include a Low Pass Filter (LPF) block and a multiplier as shown in Figure A2.1.7. The LPF block can be found in the Filters category and is the first LPF listed in the category. Recall that the MULTIPLY block outputs a 200 Hz and a 1.8 KHz sinusoid. We want to create a filter that will pass the 200 Hz and block the 1.8 KHz component. Set the LPF to have a cut-off frequency of 1 KHz and a transition width of 200 Hz. Use a Rectangular Window. Generate the flow-graph.

Remarks: After executing the flow-graph by pressing F6, students can observe that only 200 Hz component passes through the filter. Students are also advised to repeat experiment 2 using the High Pass Filter.

1.3 CONCLUSION

The basic understanding of the GRC graphic tool was developed and the use of different block in different categories was understood. In this way, different functions can be generated using blocks from different categories.

EXPERIMENT 2:   Analog Modulation and Demodulation:
Narrow Band Frequency Modulation

In this experiment, students will build a digital baseband modem that can use Narrow Band Frequency Modulation (NBFM). This will develop an understanding how practical systems evolve starting with baseband simulation to the full system implementation including transmission and reception. The main concepts in this experiment include implementation of analog modulation scheme using a baseband digital system.

This experiment comprises of two main sections. First section is the transmitting section and the second is receiving section. The overall goal of this lab is to introduce the basic concepts of analog modulation and demodulation. In this experiment, students will implement modulation and demodulation scheme for narrow band frequency modulation (NBFM) by using knowledge of building flow-graph using different building blocks as demonstrated from earlier experiment 1. Some new blocks will also be introduced to design a framework for NBFM communication link.

2.1 TRANSMITTER SIDE IMPLEMENTATION:
STEPS TO FOLLOW

Students can use following steps to develop the modulation scheme for NBFM:

Step 1: Open a terminal window using keyboard inputs: **Ctrl+Alt+T**, or by going to Dash Home on top left side and typing "Terminal" in it.

Step 2: At the terminal prompt type: **gnuradio-companion**.

Step 3: An untitled GRC window similar to the one shown in Figure A2.2.1 will open. If this window does not appear, then close all the windows and open a new window.

Step 4: Save this flow-graph.

Step 5: Double click on the Options block. This block sets some general parameters for the flow-graph. Leave the ID as top_block. Type in a project title (such as NBFM) and author. Set **Generate Options** to **QT GUI**, **Run** to **Autostart**, and **Realtime Scheduling** to **Off**. Then close the properties window.

Step 6: Open the other block named Variable block present in the flow-graph. It is used to set the sample rate value as 240,000.

Step 7: Students can find list of available blocks on the right side of the window. By expanding any of the categories (click on triangle to the left) students can see the available blocks. Students are advised to explore each of the categories to familiarize with the available blocks.

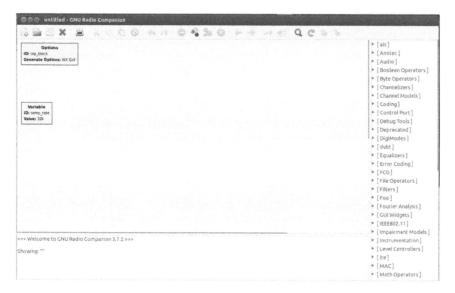

FIGURE A2.2.1
An untitled GRC window appearing in the beginning.

Step 8: Open the **Waveform Generators** category and double click on the **Signal Source**. Note that a Signal Source block will now appear in the main window. Double clicking on the block will open the properties window similar to the one shown in Figure A2.2.2. Adjust the settings as shown in Figure A2.2.2. Note that **mod_freq** is a variable which will be defined later. The signal source (in Figure A2.2.2) is now set to output a real valued sinusoid with the following characteristics:

Amplitude: 1
Frequency: mod_freq
Sample Rate: samp_rate/10 (which will be 24000)

Step 9: Open the **Modulators** category and add **NBFM Transmit** block by double clicking on it. Now an NBFM Transmit block will be added to the main window. Change the parameters to match those as shown in the Figure A2.2.3 and close the window. Following are the key parameters set:

Audio Rate: Sampling rate of incoming signal
Quadrature Rate: Sampling rate of outgoing signal (must be a integer multiple of Audio Rate)

Step 10: From the **Math Operators** category, add **Multiply Const** block. Update the values to match those in Figure A2.2.4. The amplitude of input samples to SDR should be less than ± 1, preferably even lesser than 0.8 in terms of extreme values.

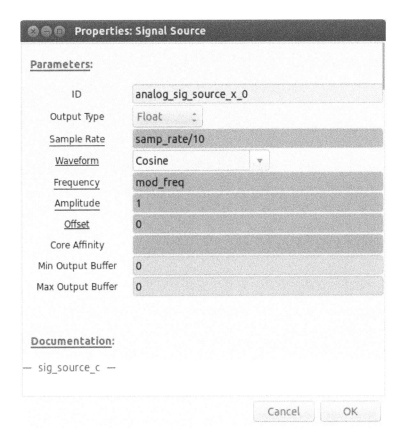

FIGURE A2.2.2
Window for setting properties of Signal Source.

Step 11: Add **QT GUI Sink**, from **Instrumentation category** and change the following parameters: **GUI Hint:** nb@0, **Bandwidth:** samp_rate as shown in Figure A2.2.5.

Step 12: Now from **GUI Widgets category**, add **QT GUI Tab Widget** from QT sub-category. Change the **ID to nb**, make two tabs and name them GUI Sink, Parameters and then click OK.

Step 13: From the category **GUI Widgets**, add **QT GUI Range** from **QT sub-category**. In the parameters change **ID to mod_freq** and all the parameters should be as shown in Figure A2.2.6. Following are the key settings in this window:
GUI Hint: nb@1

ID is like variable name which can be used as a parameter in other blocks. However, label is just a name for identifying what the variable means. It can have any valid value.

So even if one use ID as i and label as frequency, it is perfectly fine. Just update the Signal source accordingly with i in place of freq.

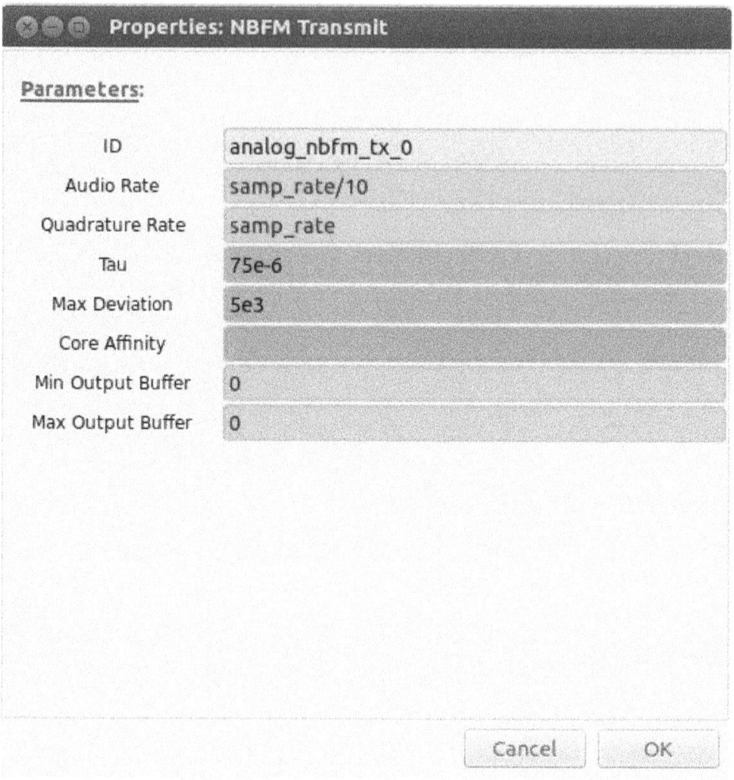

FIGURE A2.2.3
Window for setting properties of NBFM transmit block.

Step 14: From the category **GUI Widgets**, add another **QT GUI Range** from QT sub-category. In the parameters, change ID to carrier_freq and all the parameters should be as in Figure A2.2.7.

NOTE: The most common mistake here is to set invalid frequency. Use frequency in range of 300–3800 MHz, i.e., 30e6 to 3.8e9. This frequency variable will be mapped to frequency of SDR, so, it is preferable, if, different groups use different frequencies to avoid interference. Also, ensure that transmission and reception are at same frequency.

Step 15: Again from the category **GUI Widgets**, add another **QT GUI Range** from QT sub-category. In the parameters change **ID to rfgain** and all the parameters should be as in Figure A2.2.8.

Step 16: Now from the category **Amitec**, add **Amitec Sink** block and change **Device Arguments** as amitec. The parameters in Amitec Sink block should be as in Figure A2.2.9.

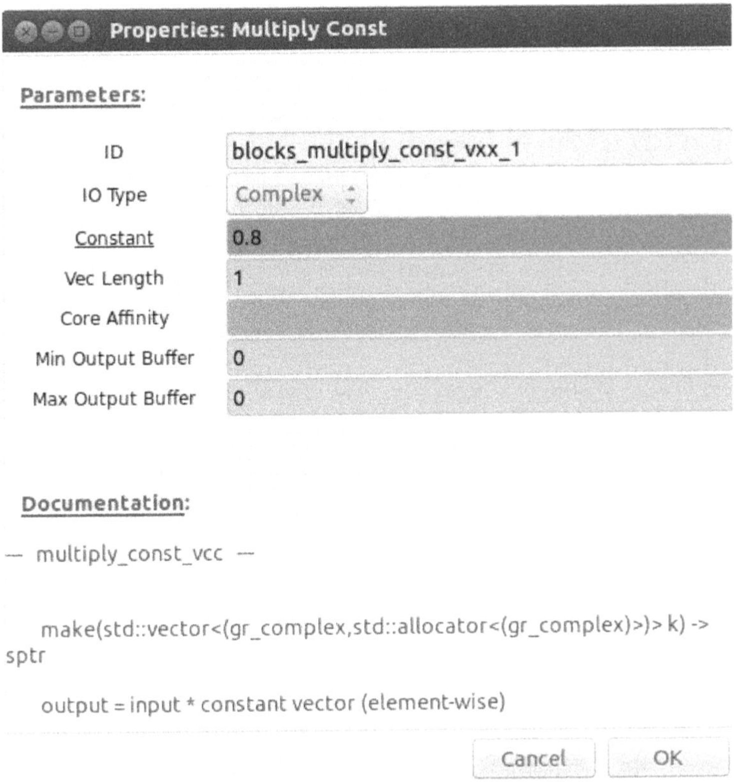

Properties: Multiply Const

Parameters:

ID	blocks_multiply_const_vxx_1
IO Type	Complex
Constant	0.8
Vec Length	1
Core Affinity	
Min Output Buffer	0
Max Output Buffer	0

Documentation:

— multiply_const_vcc —

make(std::vector<(gr_complex,std::allocator<(gr_complex)>)> k) -> sptr

output = input * constant vector (element-wise)

Cancel   OK

FIGURE A2.2.4
Window for setting properties of Multiply Const block.

Step 17: After following above steps, one can have all the blocks to complete transmitting section of NBFM modulator. Connect the blocks as shown in Figure A2.2.10.

Step 18: Now execute the flow-graph by pressing **F6**. The transmitted signal will be as shown in Figure A2.2.11.

Figure A2.2.12 shows corresponding spectrum.

One should note that there should not be any error in the terminal window.

Students task is to see the effect of changing baseband signal frequency, RF carrier frequency and rfgain with Ranges.

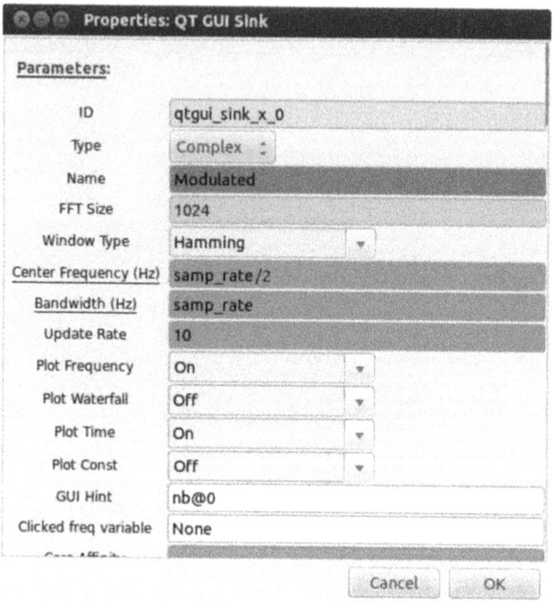

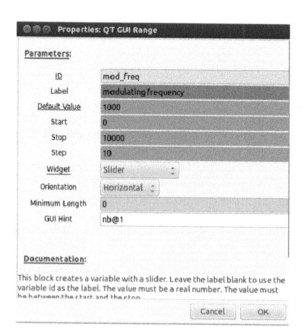

FIGURE A2.2.5
Window for setting properties of QT GUI Sink block.

FIGURE A2.2.6
Window for setting properties of QT GUI Range block.

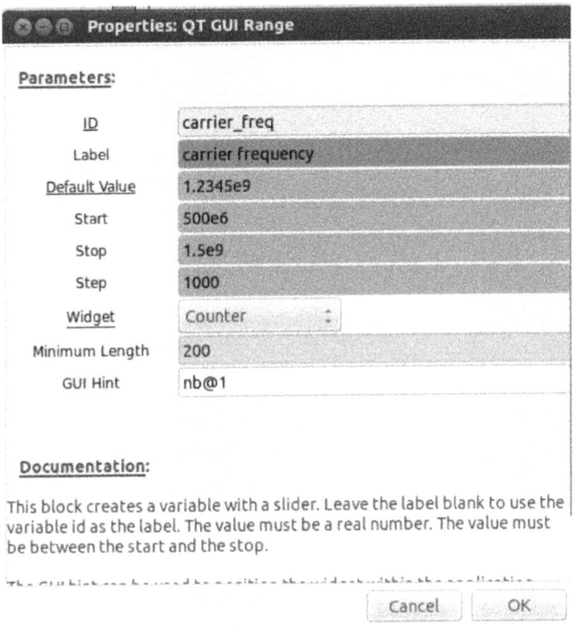

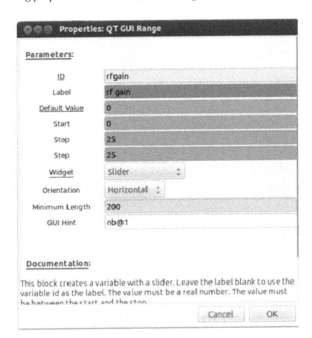

FIGURE A2.2.7
Window for setting properties of another QT GUI Range block.

FIGURE A2.2.8
Window for setting properties of third QT GUI Range block.

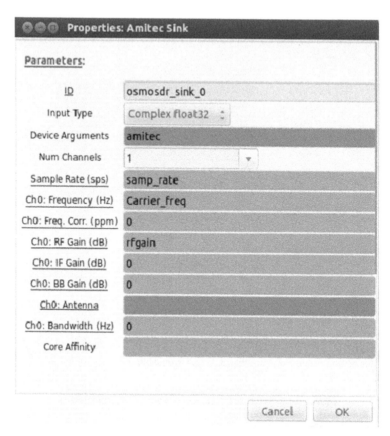

FIGURE A2.2.9
Window for setting properties of amitec sink block.

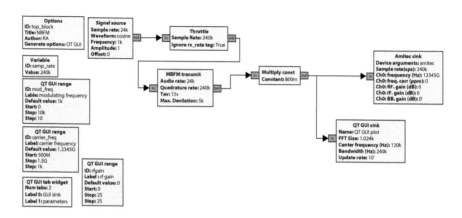

FIGURE A2.2.10
Flow-graph for transmitting section of NBFM modulator.

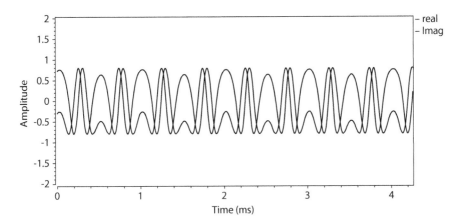

FIGURE A2.2.11
Time domain plot for NBFM modulator output.

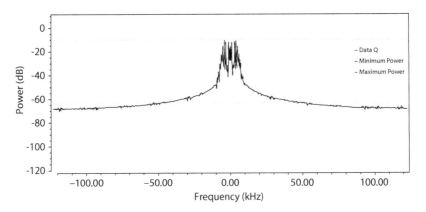

FIGURE A2.2.12
Spectrum for NBFM modulator output.

2.2 RECEIVER SIDE IMPLEMENTATION: STEPS TO FOLLOW

For receiving side implementation following steps can be used:

Step 1 to Step 4 are general steps involving starting up a window for the flow-graph development. They are same as steps 1–4 of transmitting side implementation.

Step 5: Double click on the Options block. This block sets some general parameters for the flow-graph. Set the following parameters in this block:

ID: top_block,
Project title: RX_NBFM and author,
Generate Options: QT GUI,

Run to **Autostart** and **Realtime Scheduling** to **Off**. Then close the properties window.

Step 6: Set the sampling rate of 240,000 using the variable block in the flow-graph

Step 7: On the right side of the window, there is a list of the blocks that are available. By expanding any of the categories (click on small triangle to the left of category name) one can see the different blocks available. Students are advised to explore each of the categories so that they have an idea of availability of different blocks. To search for any block, click on search button or search box can appear on the top of right side window by clicking anywhere on the block category window. Students are strongly advised to learn the types of blocks, their classification and the place where they can be find. This will save their efforts in developing future projects.

Step 8: From **GUI Widgets category**, add **QT GUI Tab Widget** from **QT sub-category**. Change the **ID** to **nb** and click **OK**. This is shown in Figure A2.2.13.

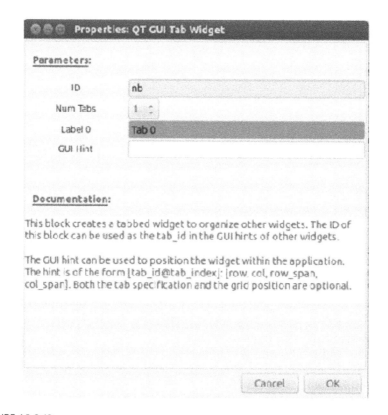

FIGURE A2.2.13
GUI window for setting properties of GUI tab widget.

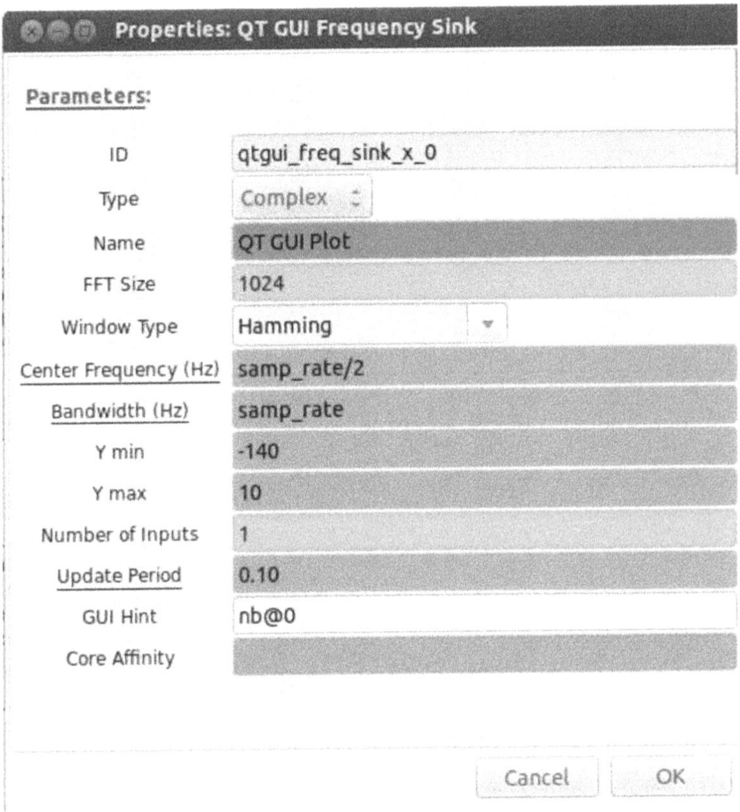

FIGURE A2.2.14
GUI window for setting frequency sink properties.

Step 9: From the **Instrumentation category**, add **QT GUI Frequency Sink** and change **GUI Hint parameter to nb@0**. This window is shown in Figure A2.2.14.

Step 10: From the category **GUI Widgets**, add another **QT GUI Range** from **QT sub-category**. Among the various parameters change **ID** to **Carrier_freq** and all the other parameters should be set as shown in Figure A2.2.15.

NOTE: The most common mistake here is to set invalid frequency. Use frequency in range of 300–3800 MHz, i.e., 30e6–3.8e9. This frequency variable will be mapped to frequency of SDR, so, it is preferable, if, different groups use different frequencies to avoid interference. Also, ensure that transmission and reception is at same frequency.

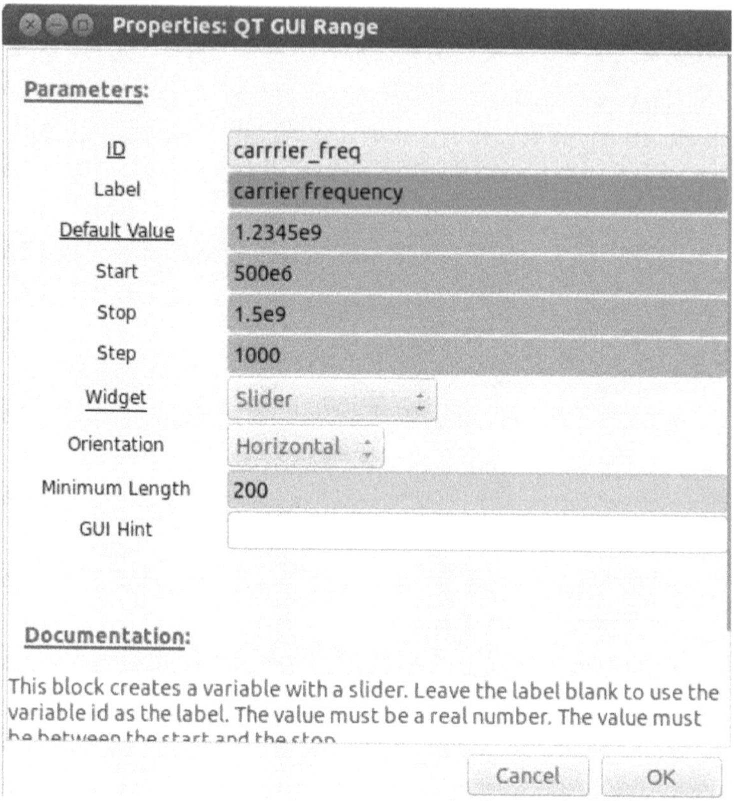

FIGURE A2.2.15
GUI window for setting QT GUI range properties.

Step 11: Now from the category **Amitec**, add **Amitec Source** and
change **Device Arguments** as **amitec** and change Parameters as
shown in Figure A2.2.16.

Step 12: Open the **Modulators** category and add **NBFM Receive**
block by double clicking on it. Now add NBFM Receive block
to the main window. Change the parameters to match those as
shown in the Figure A2.2.17 and close the window. The key settings are:

  Audio Rate: Sampling rate of outgoing signal

  Quadrature Rate: Sampling rate of incoming signal (must be
   a integer multiple of Audio Rate)

Step 13: For proper filtration at the input, students should add a
LPF. From Filters category, add LPF and adjust its parameters as
shown in Figure A2.2.18.

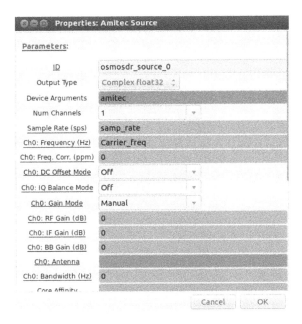

FIGURE A2.2.16
GUI window for setting Amitec Source properties.

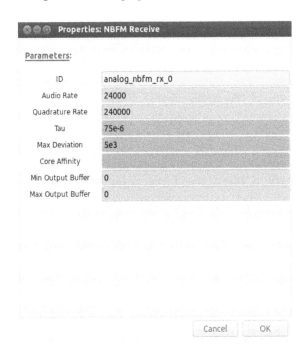

FIGURE A2.2.17
GUI window for setting NBFM receive block properties.

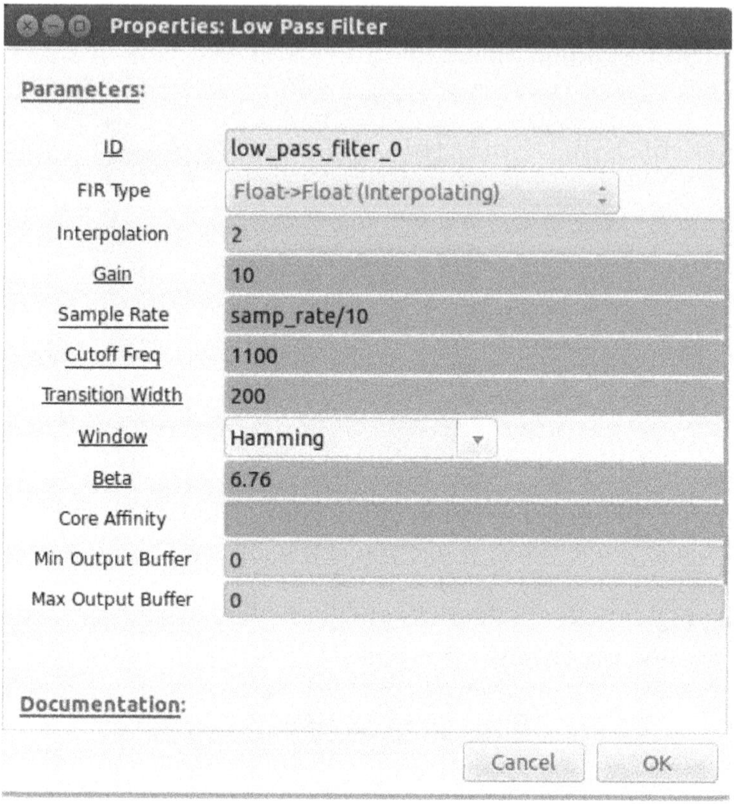

FIGURE A2.2.18
GUI window for setting LPF block properties.

Cut-off frequency and Transition width should be set according to transmitted modulating signal. By default the frequency of modulating signal is kept as 1000 Hz at the transmission end.

NOTE: The interpolation is set to 2 so that output sampling rate becomes 48000 Hz which is the sampling rate of Audio Sink

Step 14: From the **Audio** category add **Audio Sink** and change parameters as shown in Figure A2.2.19.

Audio Sink is a hardware sink, which perhaps can be the speaker of laptop running this GNU radio. It takes data at a fixed sampling rate. In this particular case the sampling rate is 48000. Many audio cards support multiple sampling rates such as 44.1 kHz, 48 kHz etc. However, 48000 Hz is supported by most of the audio card.

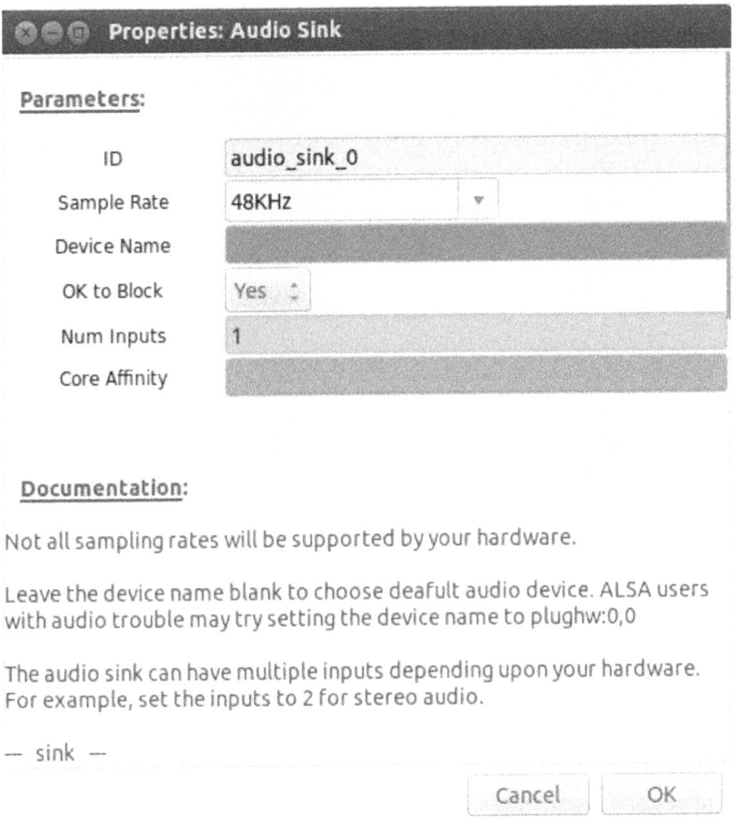

FIGURE A2.2.19
GUI window for setting Audio sink block properties.

Step 15: From **Math Operators category** add **Multiply Const block** and keep **Constant value** as 1. This can be used to adjust amplitude of samples input to audio sink.

Step 16: Following all the above steps, students have now all the blocks set in their flow-graph. Arrange all the blocks logically as shown in Figure A2.2.20.

Step 17: Execute the flow-graph of Figure A2.2.20 by pressing F6.

Step 18: The received spectrum will be displayed as shown in Figure A2.2.21.

Step 19: One can also hear a tone in the speaker. Make sure they are not set to mute.

Students are advised to change modulating frequency at transmitting end. The LPF characteristics should also be adjusted accordingly.

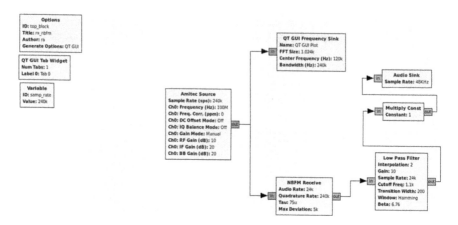

FIGURE A2.2.20
Flow-graph of NBFM receive module.

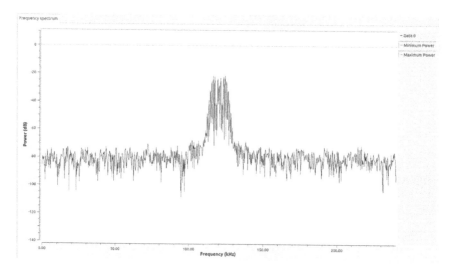

FIGURE A2.2.21
Spectrum of NBFM signal.

EXPERIMENT 3: Analog Modulation and Demodulation: Wide Band Frequency Modulation

In this lab experiment, student will build a digital baseband modem that can use Wide Band Frequency Modulation (WBFM). The basic flow of implementing the transmitting and receiving module in GRC is same as NBFM. Using knowledge of blocks from earlier experiments in conjunction with some new blocks students will design a framework for WBFM communication link.

Lab instructor can divide students into two groups: one group will implement transmitting section and another will work on receiving section.

3.1 TRANSMITTER SIDE IMPLEMENTATION: STEPS TO FOLLOW

Students can follow the following steps to develop the modulation scheme for WBFM:

Step 1: Open a terminal window using keyboard inputs: **Ctrl+Alt+T**, or by going to Dash Home on top left side and typing "Terminal" in it.

Step 2: At the terminal prompt type: **gnuradio-companion**.

Step 3: An untitled GRC window similar to the one shown in Figure A2.3.1 will open. If this window does not appear, then close all the windows and open a new window.

Step 4: Save this flow-graph.

Step 5: Double click on the Options block. This block sets some general parameters for the flow-graph as follows:
   ID: top_block.
   Project title: TX_WBFM
   Author: Student Group #, **Generate Options**: QT GUI, **Run**: Autostart,
   Scheduling: Off.
   After setting all the above parameters close the properties window.

Step 6: Open the other block named Variable block present in the flow-graph. It is used to set the **sample rate** value as 240,000.

Step 7: Students can find list of available blocks on the right side of the window. By expanding any of the categories (click on triangle to the left) students can see the available blocks. Students are advised to explore each of the categories to familiarize with the available blocks.

Step 8: Open the **Waveform Generators** category and double click on the **Signal Source**. Note that a Signal Source block will now appear in the main window. Double clicking on the block will open the properties window similar to the one shown in Figure A2.3.2. Adjust the settings as shown in Figure A2.3.2.

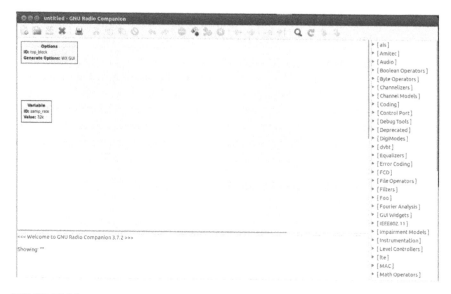

FIGURE A2.3.1

An untitled GRC window appearing in the beginning.

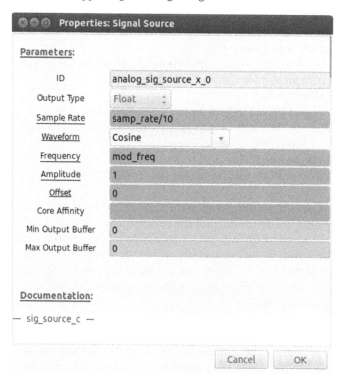

FIGURE A2.3.2

Window for setting properties of Signal Source.

Note that **mod_freq** is a variable which will be defined later. The signal source (in Figure A2.3.2) is now set to output a real valued sinusoid with the following characteristics:

    Amplitude: 1

    Frequency: mod_freq

    Sample Rate: samp_rate/10 (which will be 24000)

Step 9: Open the **Modulators** category and add **WBFM Transmit** block by double clicking on it. Now a WBFM Transmit block will be added to the main window. Change the parameters to match those as shown in the Figure A2.3.3 and close the window. Following are the key parameters set:

    Audio Rate: Sampling rate of incoming signal

    Quadrature Rate: Sampling rate of outgoing signal (must be a integer multiple of Audio Rate)

Step 10: From the **Math Operators** category, add **Multiply Const** block. Update the values to match those in Figure A2.3.4. The amplitude of samples input to SDR should be less than ± 1, preferably even lesser than 0.8 in terms of extreme values.

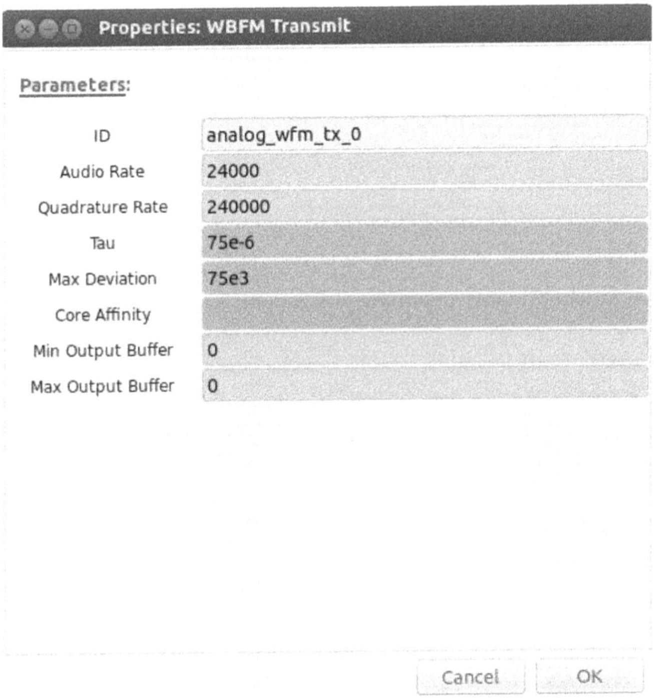

Properties: WBFM Transmit

Parameters:

ID	analog_wfm_tx_0
Audio Rate	24000
Quadrature Rate	240000
Tau	75e-6
Max Deviation	75e3
Core Affinity	
Min Output Buffer	0
Max Output Buffer	0

Cancel   OK

FIGURE A2.3.3
Window for setting properties of WBFM transmit block.

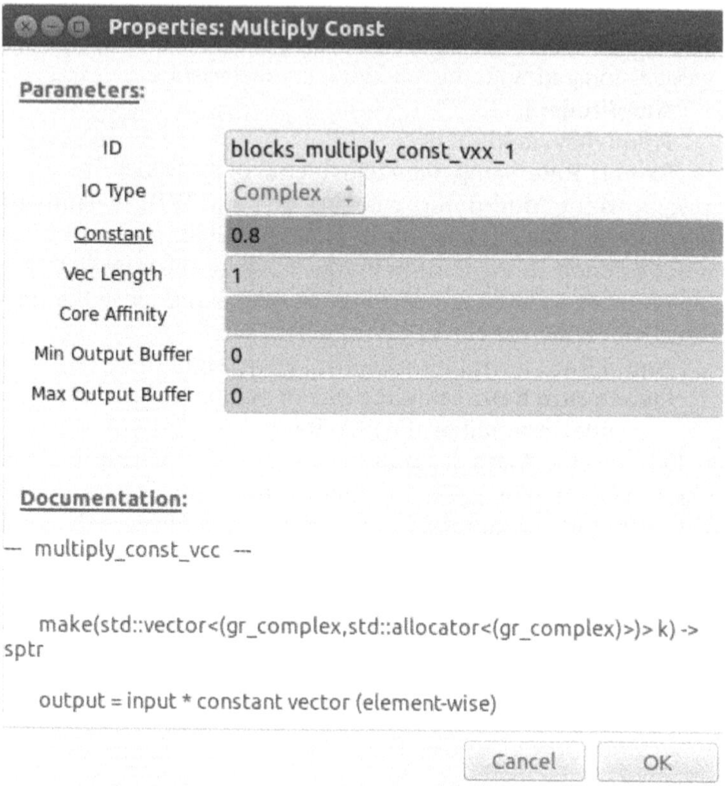

FIGURE A2.3.4
Window for setting properties of Multiply Const block.

Step 11: Add **QT GUI Frequency Sink**, from **Instrumentation cat-egory** and change the following parameters:
GUI Hint: nb@1
Bandwidth: samp_rate
Also, Add **QT GUI Time Sink**, from **Instrumentation cat-egory** and change the following parameters:
GUI Hint: nb@0
Bandwidth: samp_rate
Step 12: Now from **GUI Widgets category**, add **QT GUI Tab Widget** from QT sub-category. Change the **ID to nb**, make two tabs and name them GUI Sink, Parameters and then click OK.
Step 13: From the category **GUI Widgets**, add **QT GUI Range** from **QT sub-category**. In the parameters change **ID to mod_freq** and all the parameters should be as shown in Figure A2.3.5.

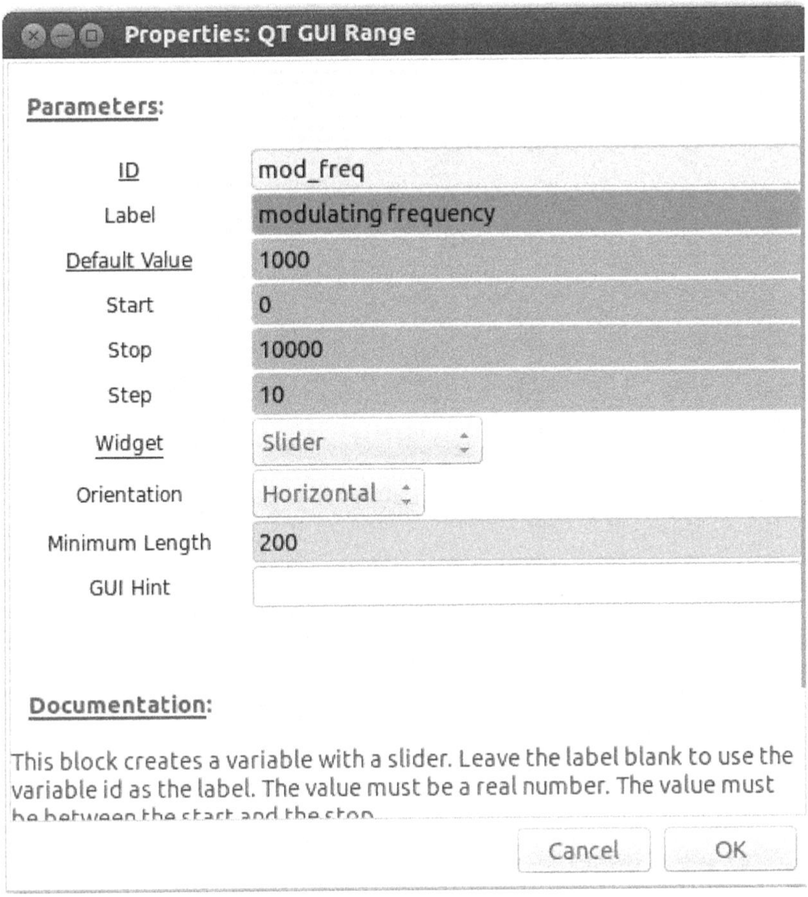

FIGURE A2.3.5
Window for setting properties of QT GUI Range block.

ID is like variable name which can be used as a parameter in other blocks. However, label is just a name for identifying what the variable means. It can have any valid value.

So even if one use ID as i and label as frequency, it is perfectly fine. Just update the Signal source accordingly with i in place of freq.

Step 14: From the category **GUI Widgets**, add another **QT GUI Range** from QT sub-category. In the parameters change ID to carrier_freq and all the parameters should be as in Figure A2.3.6.

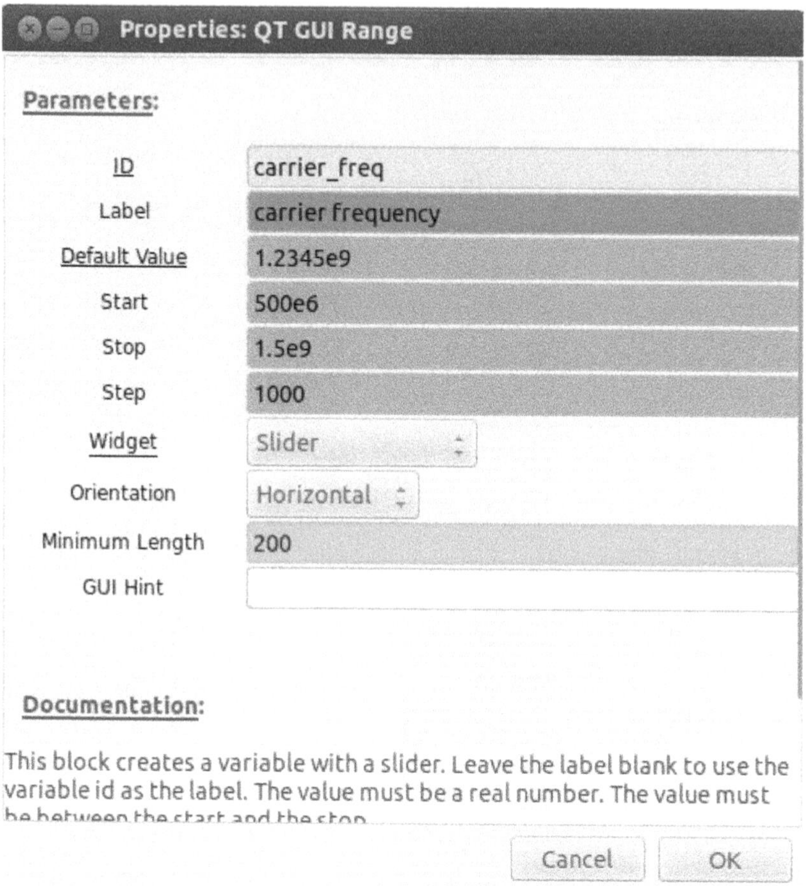

FIGURE A2.3.6
Window for setting properties of another QT GUI Range block.

NOTE: The most common mistake here is to set invalid frequency. Use frequency in range of 300–3800 MHz, i.e., 30e6–3.8e9. This frequency variable will be mapped to frequency of SDR, so, it is preferable, if, different groups use different frequencies to avoid interference. Also, ensure that transmission and reception is at same frequency.

Step 15: Again from the category **GUI Widgets**, add another **QT GUI Range** from QT sub-category. In the parameters, change **ID to rfgain** and all the parameters should be as in Figure A2.3.7.

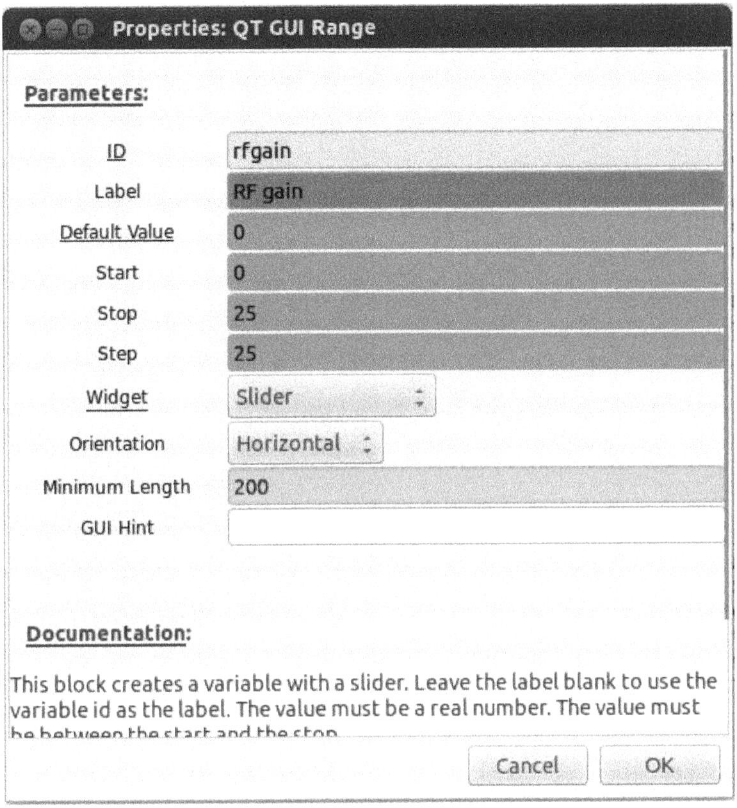

FIGURE A2.3.7
Window for setting properties of third QT GUI Range block.

Step 16: Now from the category **Amitec**, add **Amitec Sink** block and change **Device Arguments** as amitec. The parameters in Amitec Sink block should be as in Figure A2.3.8.

Step 17: After following above steps, one can have all the blocks to complete transmitting section of WBFM modulator. Connect the blocks as shown in Figure A2.3.9.

Step 18: Now execute the flow-graph by pressing **F6**. The transmitted signal will be as shown in Figure A2.3.10.

Figure A2.3.11 shows corresponding spectrum.

One should note that there should not be any error in the terminal window.

Students task is to see the effect of changing baseband signal frequency, RF carrier frequency and rfgain with Ranges.

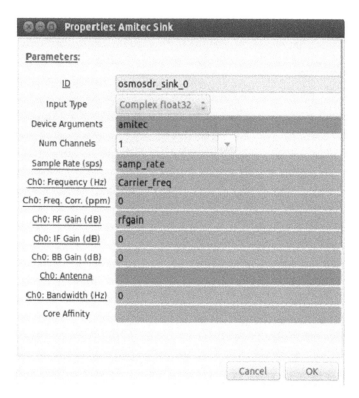

FIGURE A2.3.8
Window for setting properties of third QT GUI Range block.

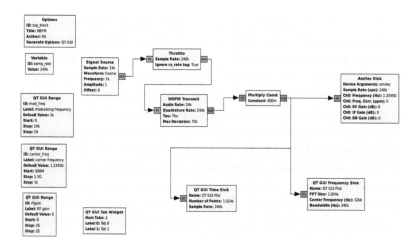

FIGURE A2.3.9
Flow-graph for transmitting section of WBFM modulator.

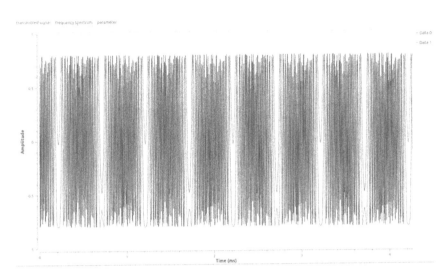

FIGURE A2.3.10
Time domain plot for WBFM modulator output.

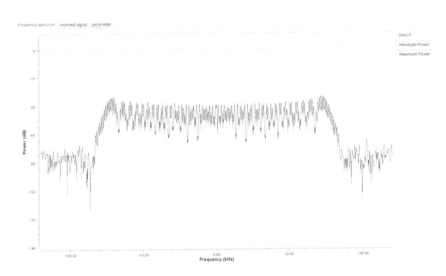

FIGURE A2.3.11
Spectrum for WBFM modulator output.

3.2 RECEIVER SIDE IMPLEMENTATION: STEPS TO FOLLOW

For receiving side implementation following steps can be used:

Step 1 to Step 4 are general steps involving starting up a window for the flow-graph development. They are same as steps 1 to 4 of transmitting side implementation.

Step 5: Double click on the Options block. This block sets some general parameters for the flow-graph. Set the following parameters in this block:

ID: top_block,
Project title: RX_WBFM and author,
Generate Options: QT GUI,
Run: Autostart
Realtime Scheduling: Off.
After setting the above parameters, close the properties window.

Step 6: Set the **sampling rate** of 240,000 using the variable block in the flow-graph

Step 7: On the right side of the window is a list of the blocks that are available. By expanding any of the categories (click on small triangle to the left of category name) one can see the different blocks available. Students are advised to explore each of the categories so that they have an idea of availability of different blocks. To search for any block click on search button or search box can appear on the top of right side window by clicking anywhere on the block category window. Students are strongly advised to learn the types of blocks, their classification and the place where they can be find. This will save their efforts in developing future projects.

Step 8: From **GUI Widgets category**, add **QT GUI Tab Widget** from **QT sub-category**. Change the **ID** to **nb** and click **OK**.

Step 9: From the **Instrumentation category**, add **QT GUI Frequency Sink** and change
GUI Hint parameter to nb@0. This window is shown in Figure A2.3.12.

Step 10: **From the category GUI Widgets,** add another **QT GUI Range** from **QT sub-category**. Among the various parameters change **ID** to **Carrier_freq** and all the other parameters should be set as shown in Figure A2.3.13.

NOTE: The most common mistake here is to set invalid frequency. Use frequency in range of 300–3800 MHz, i.e., 30e6–3.8e9. This frequency variable will be mapped to frequency of SDR, so, it is preferable, if, different groups use different frequencies to avoid interference. Also, ensure that transmission and reception is at same frequency.

Step 11: Now from the category **Amitec**, add **Amitec Source** and change **Device Arguments** as **amitec** and change Parameters as shown in Figure A2.3.14.

Step 12: Open the **Modulators** category and add **WBFM Receive** block by double clicking on it. Now add WBFM Receive block

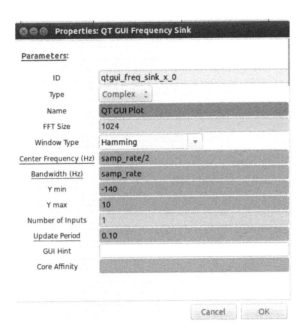

FIGURE A2.3.12
GUI window for setting frequency sink properties.

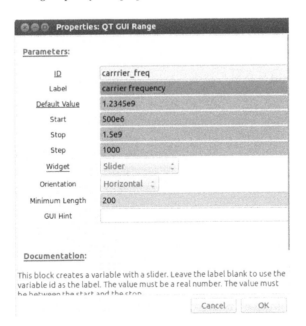

FIGURE A2.3.13
GUI window for setting QT GUI range properties.

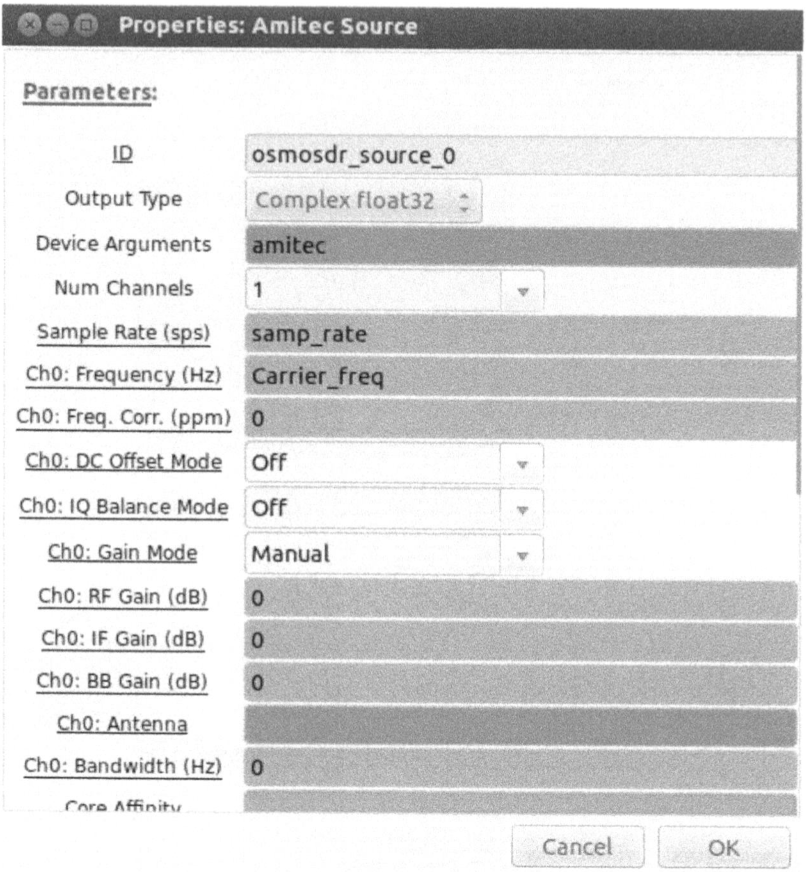

FIGURE A2.3.14
GUI window for setting Amitec Source properties.

to the main window. Change the parameters to match those as shown in the Figure A2.3.15 and close the window. The key settings are:

Audio Decimation: Factor by which incoming signal will be decimated. It should be an integer.

Quadrature Rate: Sampling rate of incoming signal (must be a integer multiple of Audio Rate)

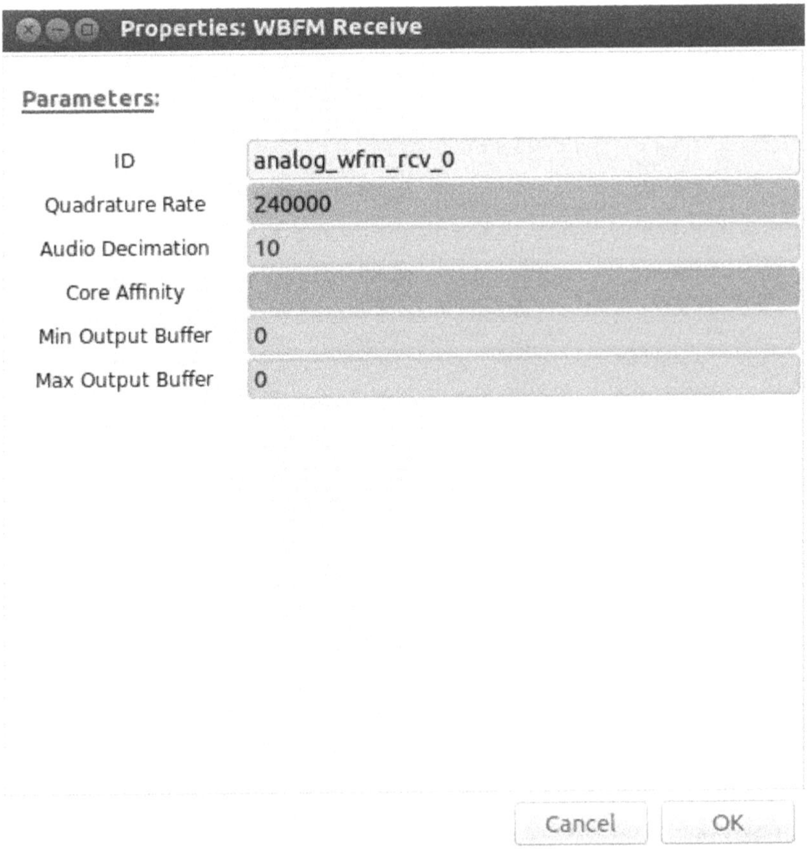

FIGURE A2.3.15
GUI window for setting WBFM receive block properties.

Step 13: For proper filtration at the input, students should add a
LPF. From Filters category, add LPF and adjust its parameters as
shown in Figure A2.3.16.

Cut-off frequency and Transition width should be set according to transmitted modulating signal. By default, the frequency
of modulating signal is kept as 1000 Hz at the transmission end.

NOTE: The interpolation is set to 2 so that output sampling rate
becomes 48000 Hz which is the sampling rate of Audio Sink

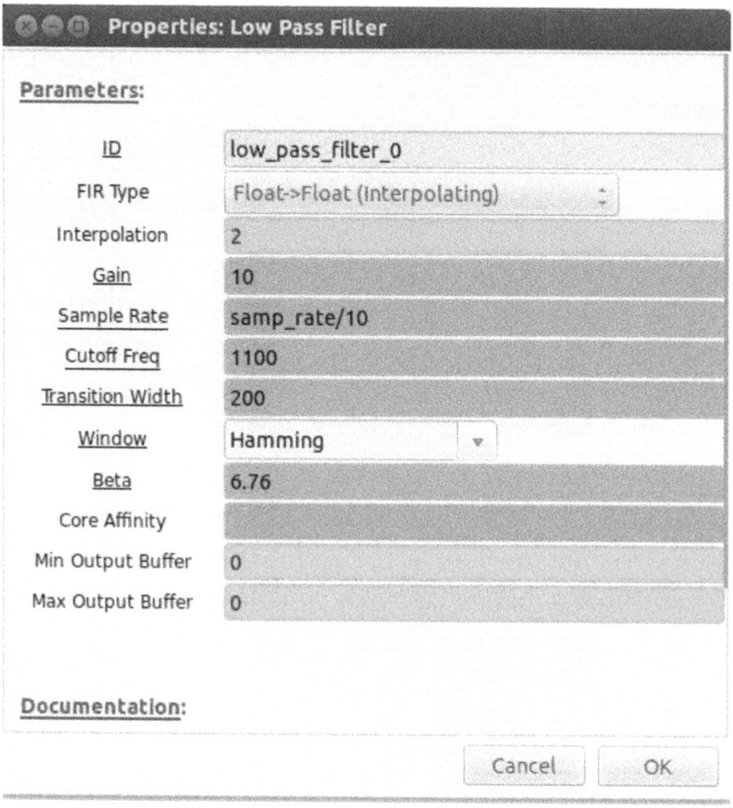

FIGURE A2.3.16
GUI window for setting LPF block properties.

Step 14: From the **Audio** category add **Audio Sink** and change parameters as shown in Figure A2.3.17.

Audio Sink is a hardware sink, which can be the speaker of laptop running this GNU radio. It takes data at a fixed sampling rate. In this particular case the sampling rate is 48000. Many audio cards support multiple sampling rates such as 44.1 kHz, 48 kHz etc. However, 48000 Hz is supported by most of the audio card.

Step 15: From **Math Operators category** add **Multiply Const block** and keep **Constant value** as 1. This can be used to adjust amplitude of samples input to audio sink.

Step 16: Following all the steps above students have now all the blocks set in their flow- graph. Arrange all the blocks logically as shown in Figure A2.3.18.

Step 17: Execute the flow-graph of Figure A2.3.18 by pressing F6.

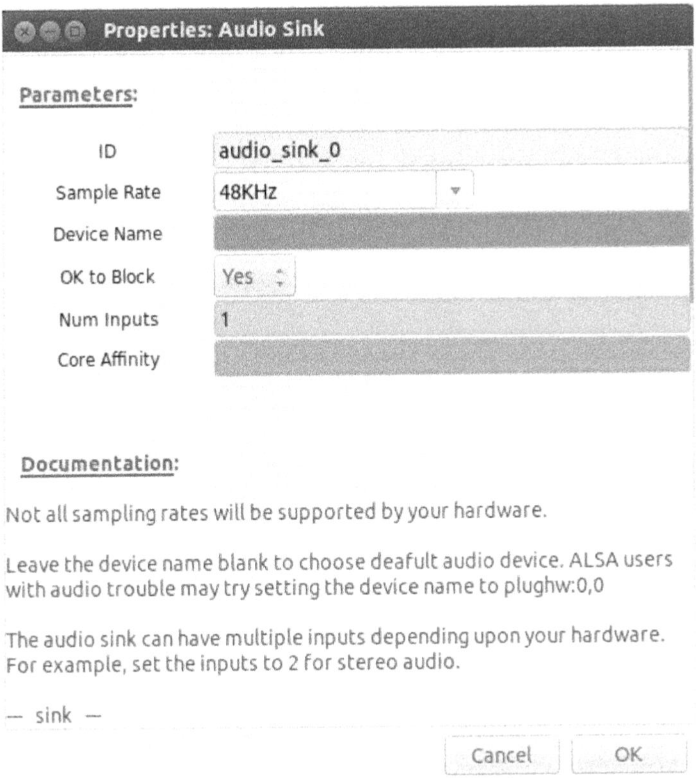

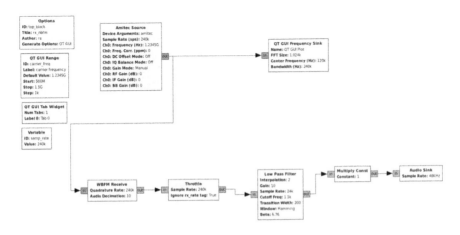

FIGURE A2.3.17
GUI window for setting Audio sink block properties.

FIGURE A2.3.18
Flow-graph of WBFM receive module.

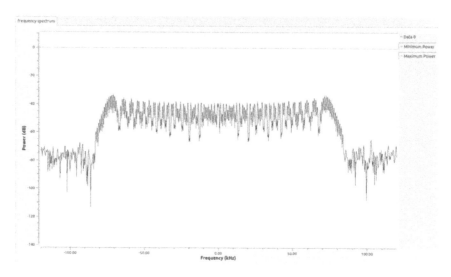

FIGURE A2.3.19
Spectrum of WBFM signal.

Step 18: The received spectrum will be displayed as shown in Figure A2.3.19.

Step 19: One can also hear a tone in the speaker. Make sure they are not set to mute.

Students are advised to change modulating frequency at transmitting end. The LPF characteristics should also be adjusted accordingly.

3.3 TROUBLESHOOTING TIPS

Following steps can be followed for the troubleshooting.

1. Check that there are no errors in terminal window and ensure that the flow-graph is actually executed. It is possible that ranges will show up even if flow-graph is not running.
2. Go through the Hardware Setup document and follow it correctly.
3. If one observes any error in terminal window saying "FATAL ERROR...," then disconnect SDR04 from machine and connect again. Wait for three LEDs to glow again.
4. Check that the flow-graph was made correctly. Even a mistake in single parameter or block will give ambiguous output which is hard to troubleshoot.
5. Most common errors are made in frequency range and sampling rate. Remember that frequency range should be between 300e6 and 3.8e9 Hz. Minimum sampling rate should be 200,000.

EXPERIMENT 4: Digital Modulation and Demodulation: Binary Phase Shift Keying Modem Implementation

This experiment targets building of digital baseband modem that can use Binary phase shift keying (BPSK). With this experiment, the students will learn basic the basic concepts of modulation and detection. This lab includes constellation mapping, bits to symbols conversion, packet encoding-decoding and symbols to bit conversion.

In this experiment, students will build the modulation and decoding blocks of the modem in GNU Radio.

4.1 BUILDING THE BPSK MODEM USING GNU RADIO COMPANION: STEPS TO FOLLOW

Students can follow the following steps to develop the modem for BPSK. The same steps can be used for transmitting as well as receiving side:

Step 1: Open a terminal window using keyboard inputs: **Ctrl+Alt+T**, or by going to Dash Home on top left side and typing "Terminal" in it.

Step 2: At the terminal prompt type: **gnuradio-companion.**

Step 3: An untitled GRC window similar to the one shown in Figure A2.4.1 will open. If this window does not appear, then close all the windows and open a new window.

Step 4: Save this flow-graph.

Step 5: Double click on the Options block. This block sets some general parameters for the flow-graph as follows:

FIGURE A2.4.1
An untitled GRC window appearing in the beginning.

ID: top_block.
Project title: BPSK
Author: Student Group #, **Generate Options**: QT GUI, **Run:**
   Autostart,
Realtime Scheduling: Off.
After setting all the above parameters close the properties
   window.
Step 6: Open the other block named Variable block present in the
   flow-graph. It is used to set the **sample rate** value as 100,000.
Step 7: Students can find list of available blocks on the right side of the
   window. By expanding any of the categories (click on triangle to the
   left) students can see the available blocks. Students are advised to
   explore each of the categories to familiarize with the available blocks.
Step 8: Open the Waveform Generators category and double click
   on the Signal Source. Note that a Signal Source block will
   now appear in the main window. Double clicking on the block
   will open the properties window similar to the one shown in
   Figure A2.4.2. Adjust the settings as shown in Figure A2.4.2.

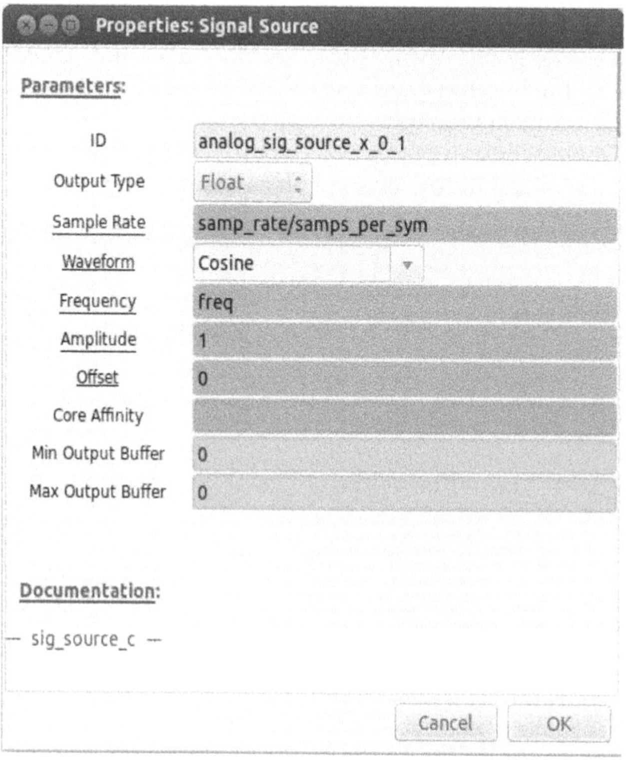

FIGURE A2.4.2
Window for setting properties of Signal Source.

Note that freq is a variable which will be defined later. The signal source (in Figure A2.4.2) is now set to output a real valued sinusoid with the following characteristics:

Amplitude: 1

Frequency: freq

Sample Rate: samp_rate/samps_per_sym

Step 9: Open the **Misc** category and add **Throttle** block. The Throttle block is needed because simulation will be carried out in this part. Set the following parameter values:

Sample Rate parameter: samp_rate/samps_per_sym.

samp_rate variable is already defined as 100,000 and samps_per_sym variable will be defined later.

The throttle block will limit the rate at which data is passed between blocks. If throttle block is not used then complete CPU processing will be used and system may hang. Open the throttle block and change **Type** as **Byte**.

Step 10: Open the **Modulators** category and add **PSK Mod** block by double clicking on it. Once a PSK Mod block is added to the main window, change the parameters to match those as shown in the Figure A2.4.3 and close the window.

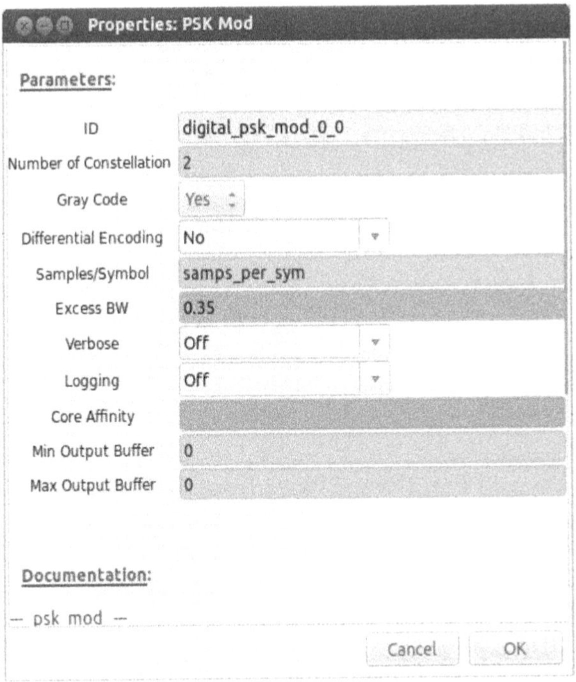

FIGURE A2.4.3
Window for setting properties of PSK Mod block.

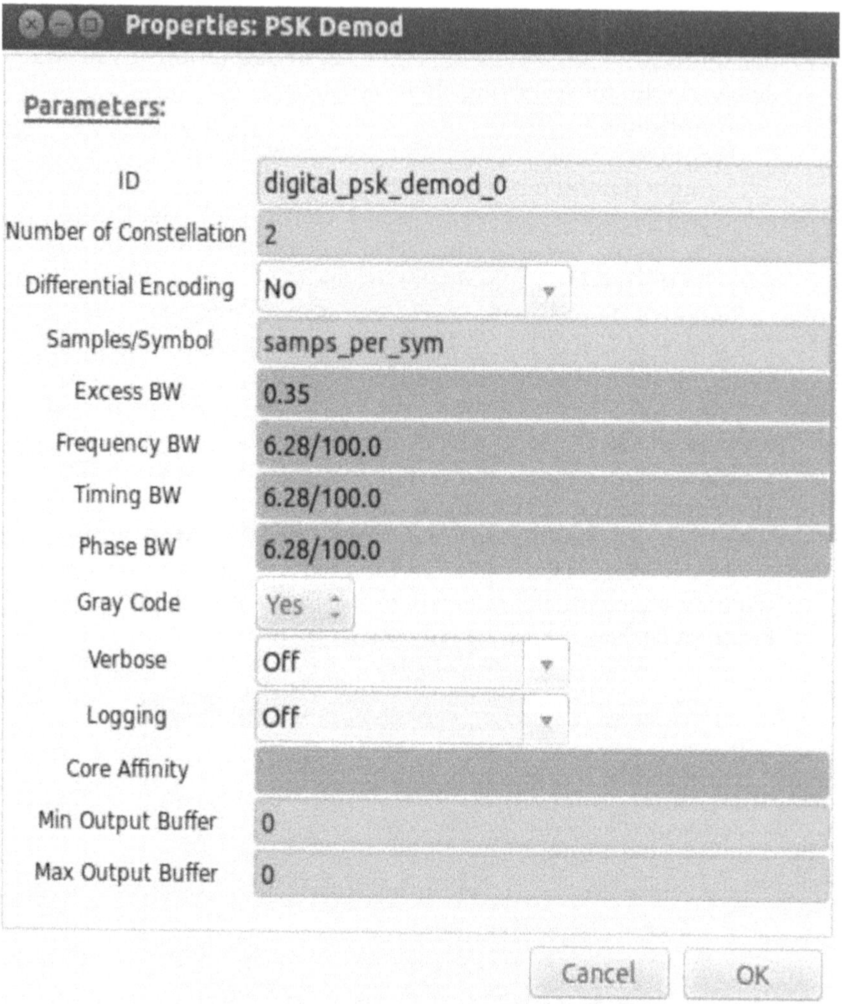

⊗ ⊖ ⊡ Properties: PSK Demod

Parameters:

ID	digital_psk_demod_0
Number of Constellation	2
Differential Encoding	No ▾
Samples/Symbol	samps_per_sym
Excess BW	0.35
Frequency BW	6.28/100.0
Timing BW	6.28/100.0
Phase BW	6.28/100.0
Gray Code	Yes ⇕
Verbose	Off ▾
Logging	Off ▾
Core Affinity	
Min Output Buffer	0
Max Output Buffer	0

Cancel   OK

FIGURE A2.4.4
Window for setting properties of PSK Demod block.

Step 11: Now add PSK Demod block from the Modulators category and change its properties as shown Figure A2.4.4.

Step 12: Now, one need to add packet information to define the start and stop of each packet that will be modulated and demodulated. This can be done by using Packet Encoder block at the transmitting side and correspondingly using Packet Decoder block at the receiving side.

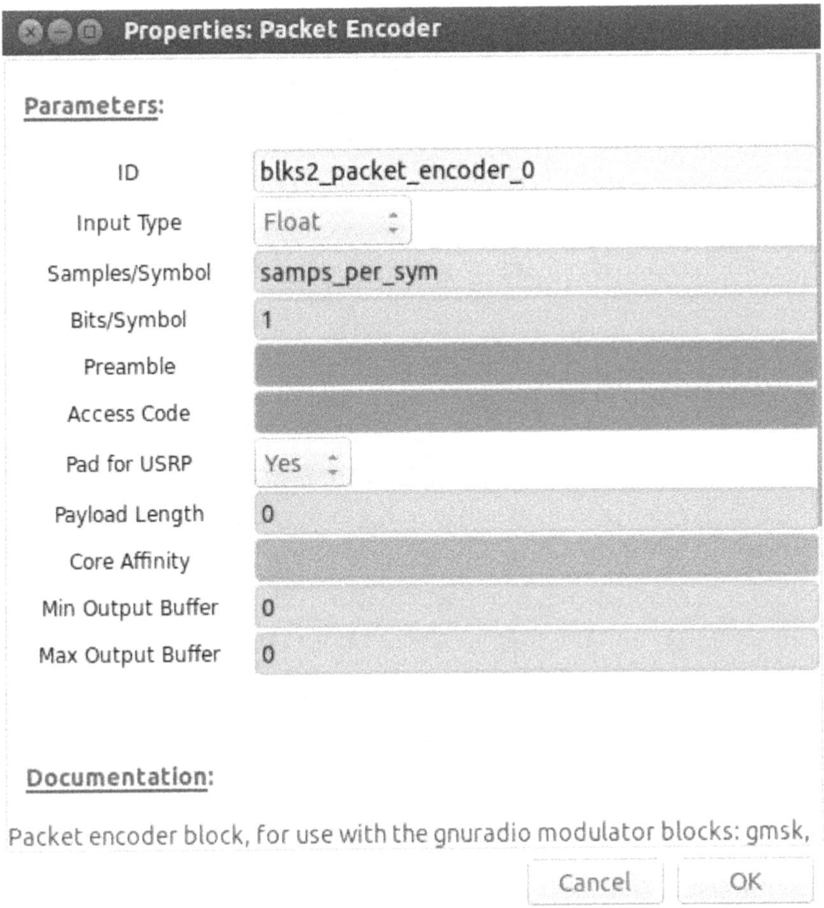

Properties: Packet Encoder

Parameters:

ID	blks2_packet_encoder_0
Input Type	Float
Samples/Symbol	samps_per_sym
Bits/Symbol	1
Preamble	
Access Code	
Pad for USRP	Yes
Payload Length	0
Core Affinity	
Min Output Buffer	0
Max Output Buffer	0

Documentation:

Packet encoder block, for use with the gnuradio modulator blocks: gmsk,

Cancel   OK

FIGURE A2.4.5
Window for setting properties of Packet encoder block.

From the Packet Operators category, add Packet Encoder and Packet Decoder blocks. There is no need to change any parameter in the Packet Decoder block. However, change the properties of Packet Encoder block to match those as shown in Figure A2.4.5.
Step 13: Add Variable block from the Variables category. Enter the parameters of Variable block as:

    ID: samps_per_sym
    Value: 2

This will appear as shown in Figure A2.4.6.

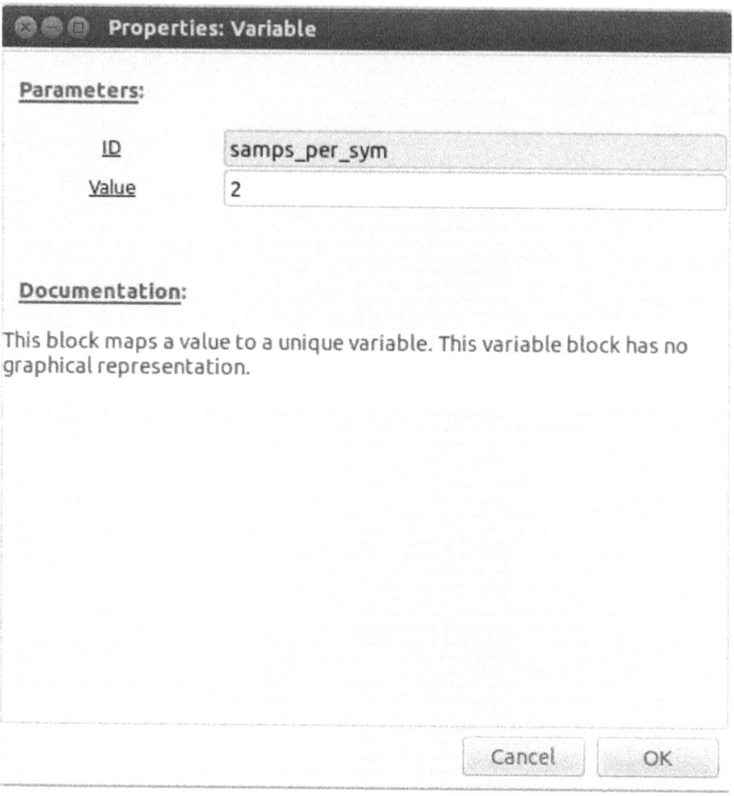

FIGURE A2.4.6
Window for setting properties of Variable block.

Step 14: To visualize the output waveform, add **QT GUI Time Sink** from **Instrumentation category**, which is under **QT category**. Change the parameters of sink to those as shown in Figure A2.4.7. Change the Type to Float and Num Inputs to 2.

Step 15: Following all the above steps, students have placed all the components to create a basic BPSK simulation flow-graph. Connect them to create a flow-graph as shown in Figure A2.4.8.

Step 16: Execute the flow-graph by pressing F6.

NOTE: The PSK Demod implementation utilizes several blocks and to observe the output at each of these sub-blocks one need to discretely implement the PSK Demod block. This will be done in later labs. However, to just observe the constellation, one more block is needed.

Step 17: After playing with the parameters, bring the flow-graph back to default state.

Step 18: From **Synchronizers** category, add **Polyphase Clock Sync** and change the parameters of match as shown in Figure A2.4.9.

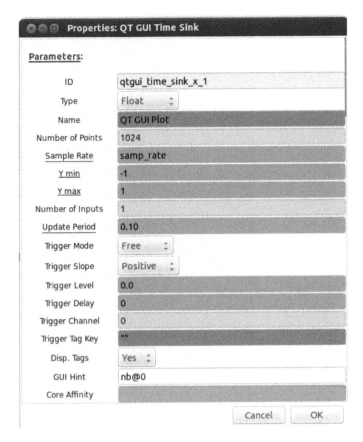

FIGURE A2.4.7
Window for setting properties of Time sink block.

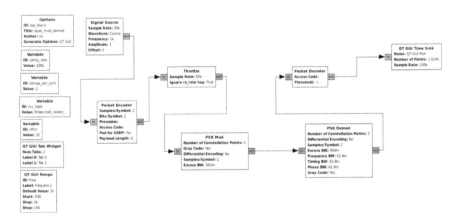

FIGURE A2.4.8
Flow-graph for BPSK simulation.

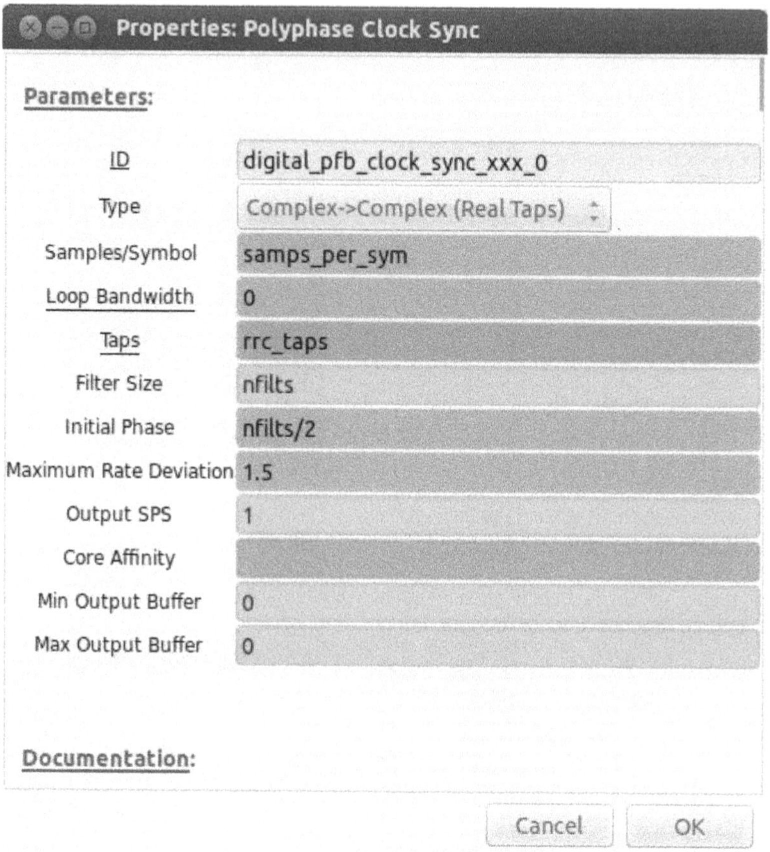

FIGURE A2.4.9
Window for setting properties of Polyphase clock Sync.

Step 19: From the Variables category, add another Variable block. Enter the parameters of Variable block as:
    ID: nfilts
    Value: 32
Step 20: From the Variables category, add Variable block. Enter the parameters of Variable block as:
    ID: rrc_taps
    Value: firdes.root_raised_cosine(nfilts, nfilts,1.0/float(samps_per_sym),0.35,11*samps_per_sym*nfilts)
    firdes: is GNU Radio filter design utility
Step 21: Now from **GUI Widgets category**, add **QT GUI Tab Widget** from **QT sub-category**. Change the **ID** to **nb** and click OK.
Step 22: Open the Time Sink added in Step 14. Add the value of parameter **GUI Hint** as **nb@0**.

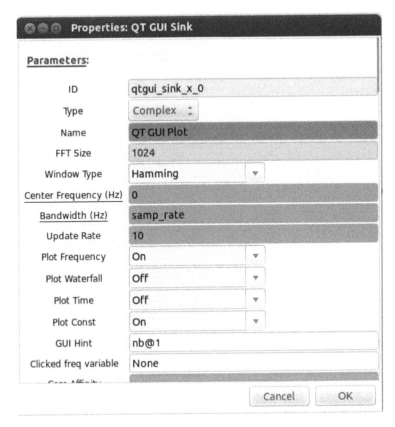

FIGURE A2.4.10
Window for setting properties of QT GUI Sink.

Step 23: Add a new **QT GUI Sink** from **Instrumentation category** and switch **the Plot Frequency** and **Plot Const ON** and add **GUI Hint value** as **nb@1**. This is shown in Figure A2.4.10.

Step 24: Connect all the blocks added to complete the flow-graph as shown in Figure A2.4.11.

Step 25: Now open Poly-phase Clock Sync block and change **Loop Bandwidth** to **0.01**.

Step 26: Execute the flow-graph and observe the scope in Tab 1. The constellation plot will look similar to the one shown in Figure A2.4.12. This is a BPSK constellation. After pulse shaping is done in PSK Mod block, the subsequent blocks, i.e., Polyphase Clock Sync block and Scope sink in this case continuously sample the incoming data. Polyphase Clock Sync block has compensated for any scattering in constellation plot this. Save this flow-graph as it will be used in next labs.

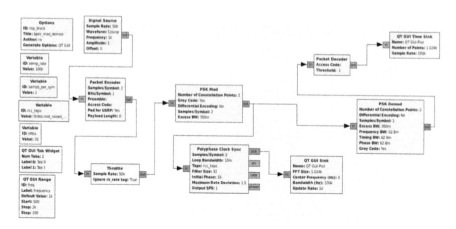

FIGURE A2.4.11
Flow-graph for BPSK simulation with feature of constellation plot.

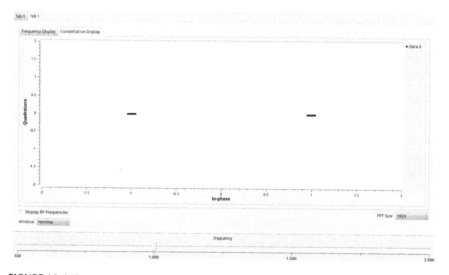

FIGURE A2.4.12
Constellation plot for BPSK modulation.

EXPERIMENT 5: Digital Communication Based on Binary Phase Shift Keying Modem Using RF Communication Link

In this experiment, students will build a digital over the air modem that can use Binary Phase Shift Keying (BPSK) modulation. In Experiment 4, students have already created a baseband BPSK modem. Taking that flow-graph as guideline, students will create a RF link in this experiment. In earlier experiment, the baseband simulation was ideal and involves the working of a modem without much deviation from theory. However, in real time transmission and reception, many additional parameters should be taken into account. This experiment includes introduction to frequency offset correction, phase synchronization and timing offset correction required in transmission over real RF link.

In this experiment, the modem transmits symbols and then demodulates symbols over the air. The transmitter maps bits to elements of a symbol constellation. The sequence of symbols are then filtered in the discrete-time by a pulse shaping filter. These symbols are then given as input to Amitec Sink which controls transmission part of SDR04. The data is received over the air and the receiving section dumps the data through Amitec Source block. This data is then taken from output port of Amitec Source and can be used in subsequent blocks. The receiving blocks sample the received signal and filter it with the receiver pulse shaping filter. The filtered symbols are then inversely mapped to produce bits. Following sections describe steps to be followed for using GNU Radio Companion to build BPSK modem on SDR04.

Lab instructor can divide students into two groups: one group will implement transmitting section and another will work on receiving section.

5.1 TRANSMITTER SIDE IMPLEMENTATION: STEPS TO FOLLOW

Students can use the following steps to develop the BPSK modulation in the transmitter side:

Step 1: Open a terminal window using keyboard inputs: **Ctrl+Alt+T**, or by going to Dash Home on top left side and typing "Terminal" in it.

Step 2: At the terminal prompt type: **gnuradio-companion**.

Step 3: An untitled GRC window similar to the one shown in Figure A2.5.1 will open. If this window does not appear, then close all the windows and open a new window.

Step 4: Open the flow-graph created in Figure A2.4.11 of previous experiment. This will be used as starting point.

Step 5: Remove the blocks Throttle, PSK Demod, Packet Decoder, Polyphase Clock Sync and both the Scope sinks from the flow-graph opened in step 4.

Step 6: Open the Variable **samp_rate** and change value to **200,000**.

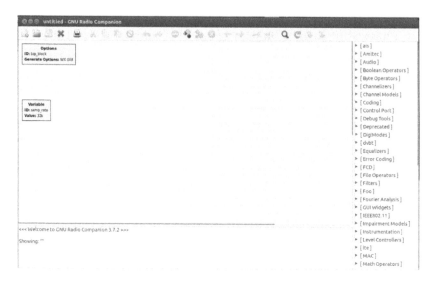

FIGURE A2.5.1
An untitled GRC window appearing in the beginning.

Step 7: Open the Variable **samps_per_sym** and change value to **4**.

Step 8: Open the block **Signal Source** and change the following:

Sample Rate: samp_rate/samps_per_sym

Frequency: freq.

Frequency will be a parameter which we will assign to a Range so that it can be changed at run-time.

Click OK and Sample Rate should be updated to 50000. The updated Signal Source block should appear as shown in Figure A2.5.2.

Step 9: From the category **GUI Widgets**, add **QT GUI Range** from **QT sub-category**. In the parameters, change **ID** to **freq** and all the other parameters should remain as shown in Figure A2.5.3.

Step 10: From the category **GUI Widgets**, add another **QT GUI Range** from **QT sub-category**. Among the parameters change **ID** to **Carrier_freq** and keep all the parameters as shown in Figure A2.5.4.

Step 11: Again from the category **GUI Widgets**, add another **QT GUI Range** from **QT sub-category**. Among the parameters change **ID** to **rfgain** and all the parameters should shown in be set as Figure A2.5.5.

Step 12: From the category **Math Operators**, add block **Multiply Const**. Keep the **IO Type** as **Complex** and **Constant value** to 0.3.

Step 13: From the category **Amitec**, add **Amitec Sink** block and change **Device Arguments** as **amitec**. The parameters in Amitec Sink block should be set as Figure A2.5.6.

Step 14: Connect all the blocks added in previous steps to complete transmitting section of BPSK modulator. The overall flow-graph is shown Figure A2.5.7.

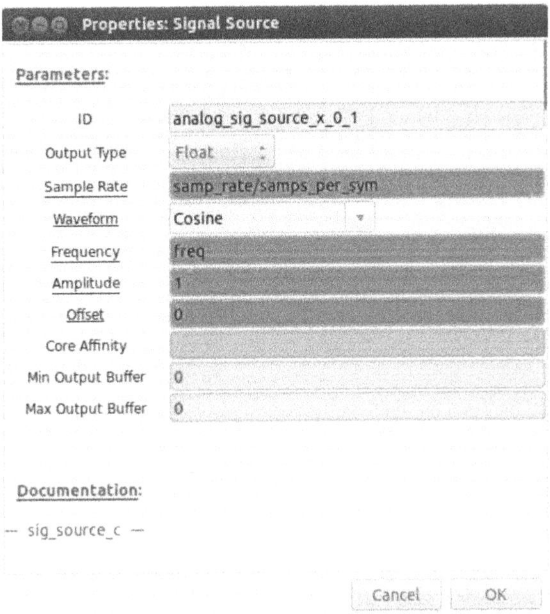

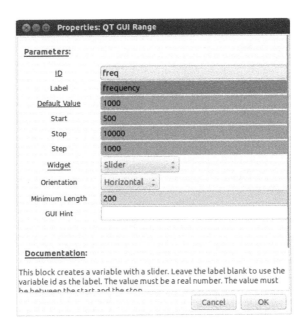

FIGURE A2.5.2
Window showing the updated Signal Source block.

FIGURE A2.5.3
Window showing the QT GUI range block.

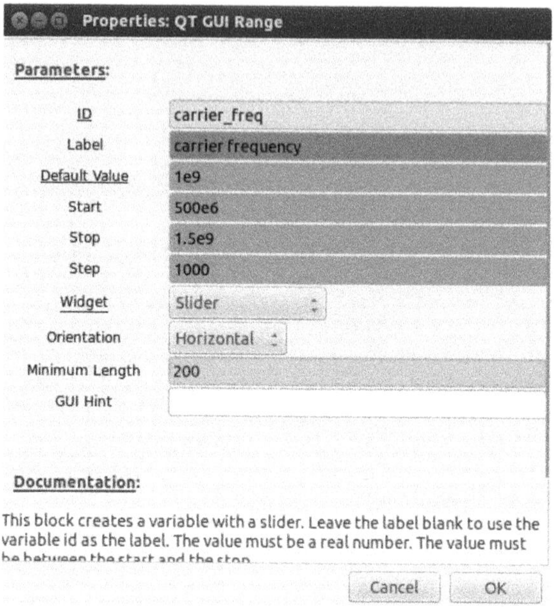

FIGURE A2.5.4
Window showing second QT GUI range block.

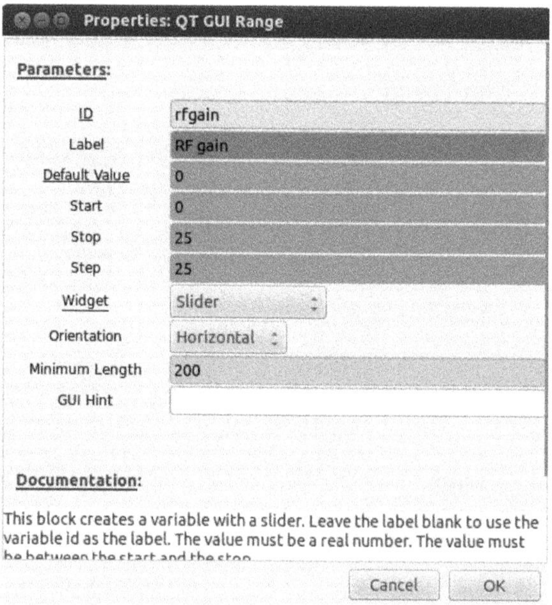

FIGURE A2.5.5
Window showing third QT GUI range block.

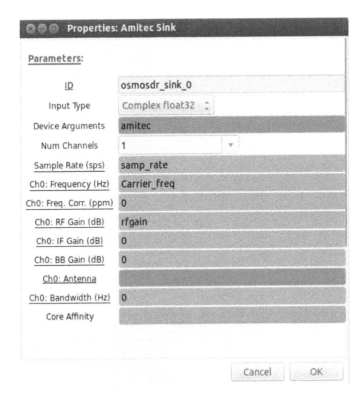

FIGURE A2.5.6
Window showing updated Amitec sink.

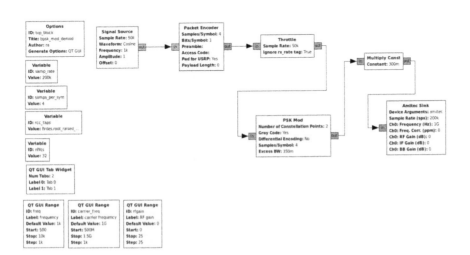

FIGURE A2.5.7
Flow-graph of BPSK modulator.

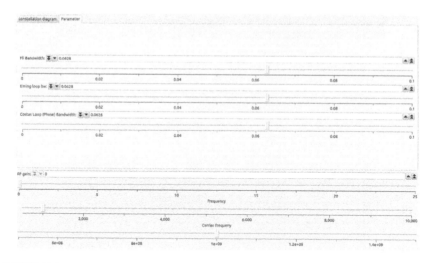

FIGURE A2.5.8
Window for setting parameters for constellation plot.

```
Executing: "/home/kartik/Dropbox/SDR/GNURadio_files/Transceivers/Transmitter/Fin
al/top_block.py"

linux; GNU C++ version 4.6.3; Boost_104800; UHD_003.005.002-0-g0cce80c1

gr-osmosdr v0.1.0-44-g0d10f5e9 (0.1.1git) gnuradio 3.7.2
built-in sink types: uhd amitec
[INFO] Instance: 0
[amitec source] Using Amitec SDR40 #0 SN b9de749fdc17763765502055f4420470 FW v1.
5.3 FPGA v0.0.0
Using Volk machine: avx_64_mmx_orc
```

FIGURE A2.5.9
Terminal Window for checking errors.

> Step 15: Now execute this flow-graph by pressing F6. Three Ranges will be created as shown in Figure A2.5.8.
> Step 16: Double check the terminal window as shown in Figure A2.5.9 for any errors.
> Step 17: You can also change baseband signal frequency, RF carrier frequency and rfgain with Ranges.

5.2 RECEIVER SIDE IMPLEMENTATION: STEPS TO FOLLOW

Students can follow the following steps to develop the BPSK demodulation in the receiver side:

> Step 1: Open the flow-graph created in Figure A2.4.11 of previous experiment. This will be used as starting point. Remove the blocks Throttle, PSK Mod, Packet Encoder, from this flow-graph.
> Step 2: Open the Variable **samp_rate** and change value to **200,000**.
> Step 3: Open the Variable **samps_per_sym** and change value to **4**.

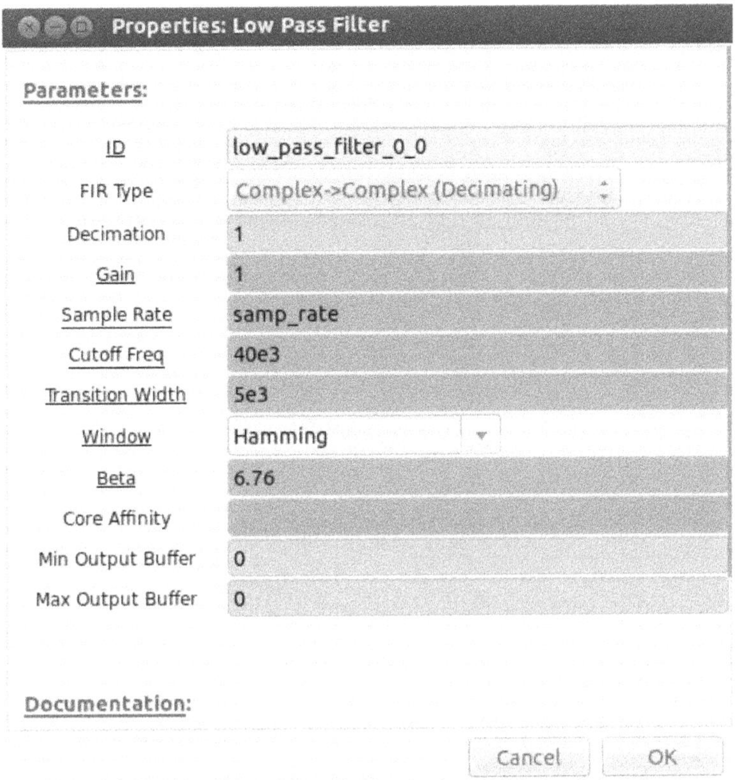

FIGURE A2.5.10
Window for setting LPF parameters.

Step 4: Add a Low pass filter for proper filtering at the input. From Filters category, add LPF and adjust its parameters to match as Figure A2.5.10.

Step 5: From the category **GUI Widgets**, add **QT GUI Range** from **QT sub-category**. Among the parameters change **ID** to **freq_bw** and all other parameters should be set as shown in Figure A2.5.11.

Step 6: Again from the category **GUI Widgets**, add another **QT GUI Range** from **QT sub-category**. Among the parameters change **ID** to **time_bw** and all other parameters should be set as shown in Figure A2.5.12.

Step 7: Add one more **QT GUI Range** from **category GUI Widgets** and **QT sub-category**. Among the parameters change **ID** to **phase_bw** and set all other parameters as shown in Figure A2.5.13.

Step 8: From the **Synchronizers category**, add **FLL Band-Edge block** and change its parameters as shown in Figure A2.5.14. This block will be used for frequency offset correction.

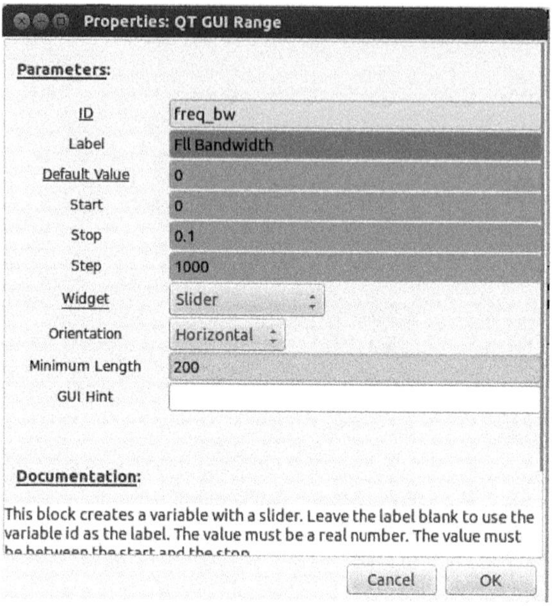

FIGURE A2.5.11
Window for setting frequency range.

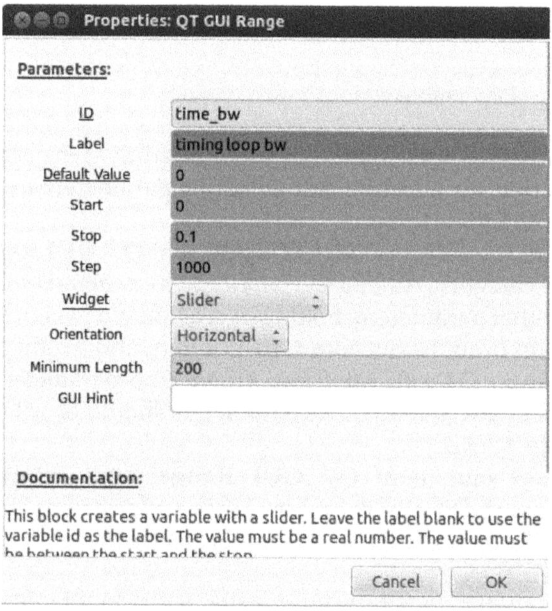

FIGURE A2.5.12
Window for setting time range.

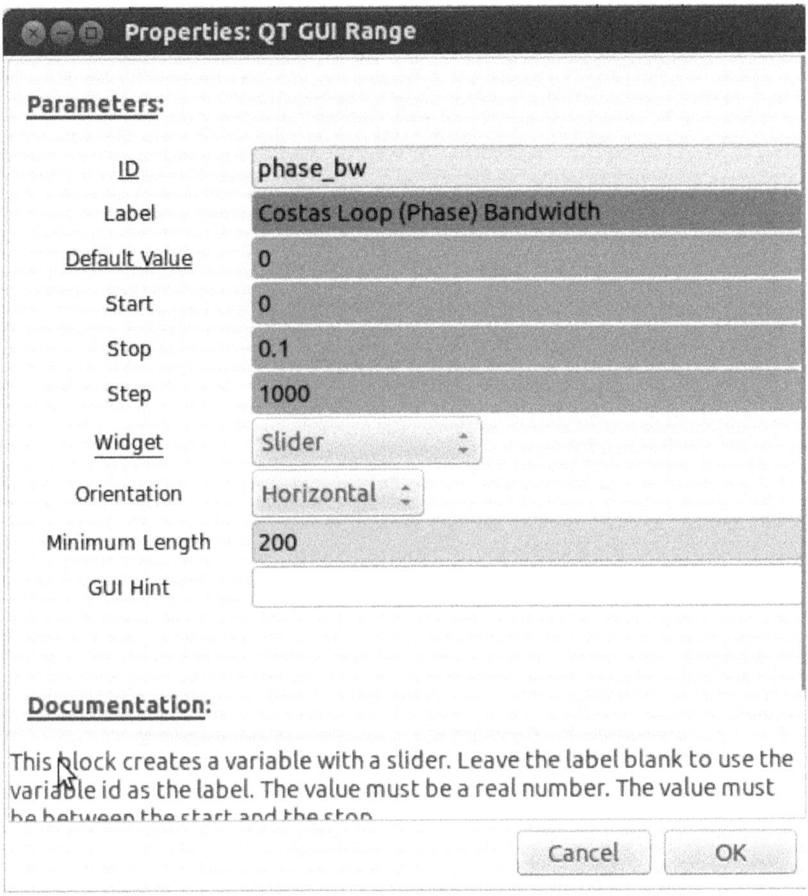

FIGURE A2.5.13
Window for setting phase range.

Step 9: From the **Synchronizers category**, add **Polyphase Clock Sync block** and change parameters as shown in Figure A2.5.15. This block will be used for timing synchronization.

Step 10: Again from the Synchronizers category add Costas Loop and set its parameters as shown in Figure A2.5.16. This block will be used for phase offset correction.

Step 11: From the **Instrumentation category**, add **QT GUI Frequency Sink** and change **GUI Hint parameter** to **nb@1.**

Step 12: Ensure that QT GUI Tab Widget block has 3 tabs.

Step 13: Open the Time Sink connected to Packet Decoder. Make the changes as per Figure A2.5.17.

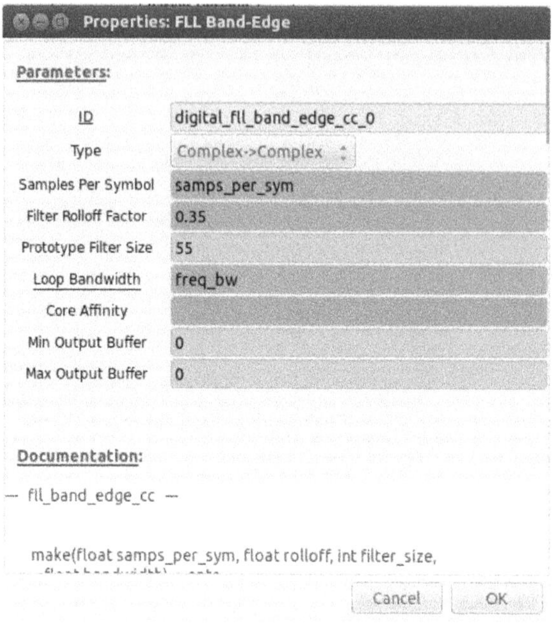

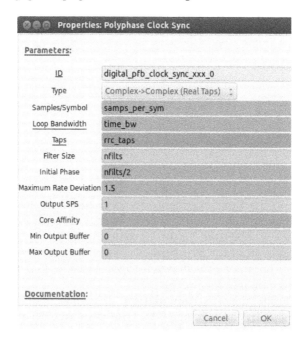

FIGURE A2.5.14
Window showing updated properties of FLL Band edge block.

FIGURE A2.5.15
Window showing updated properties of Polyphase Clock Sync block.

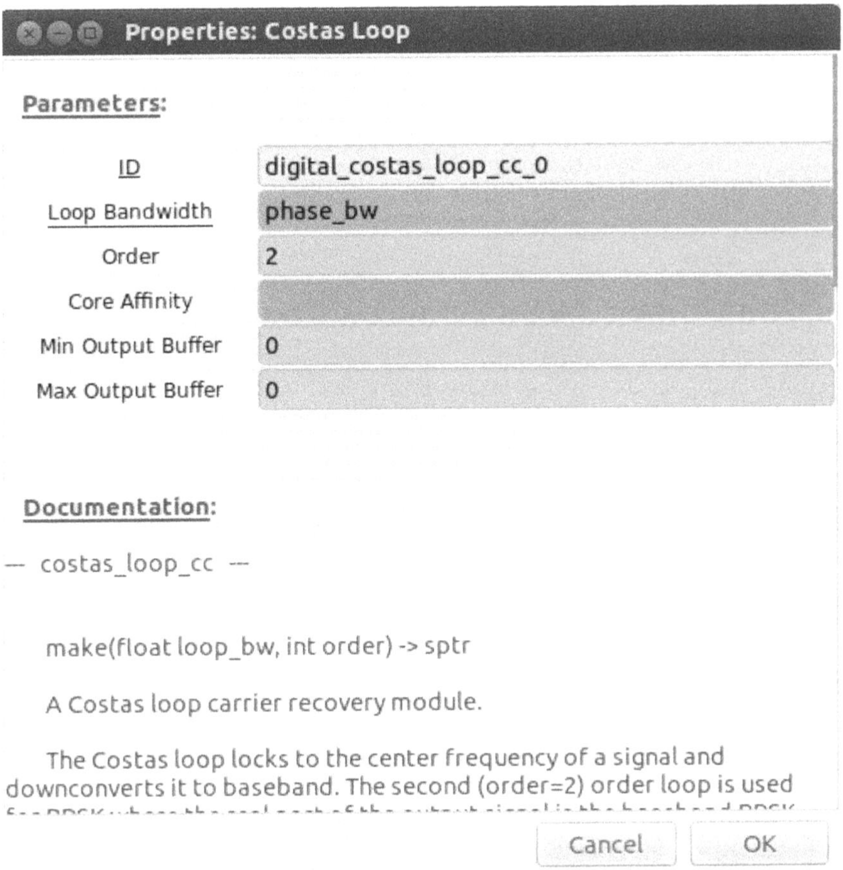

FIGURE A2.5.16
Window showing updated properties of Costas loop block.

Step 14: Open the other QT GUI Sink in flow-graph and change **Sample Rate** to **samp_rate/samps_per_sym** and **GUI Hint** to nb@0 as in Figure A2.5.18.

Step 15: From the category **GUI Widgets**, add another **QT GUI Range** from **QT sub-category**. Among the parameters, change **ID** to **Carrier_freq** and all the parameters are set as shown in Figure A2.5.19.

NOTE: Different groups should use different Default value and preferably separate bands which can be controlled by Minimum and Maximum. Otherwise there might be interference issues.

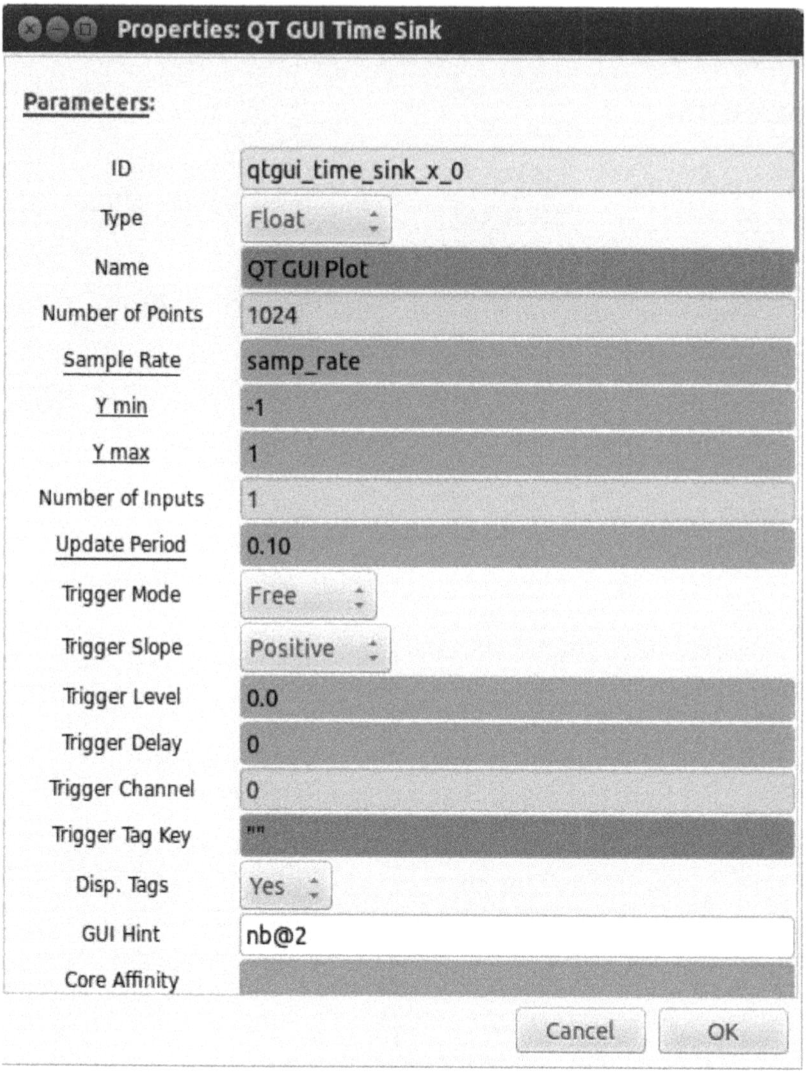

FIGURE A2.5.17
Window showing updated properties of Time sink block.

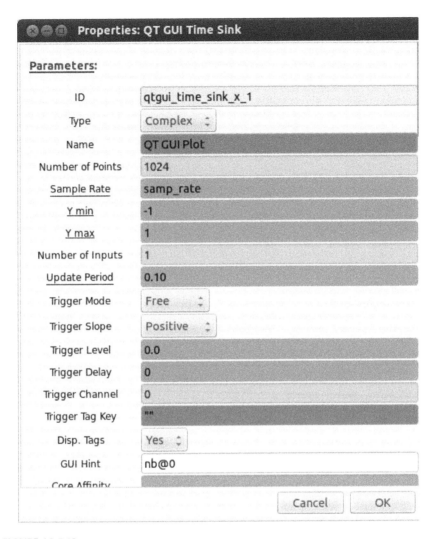

FIGURE A2.5.18
Window showing updated properties of second Time sink block.

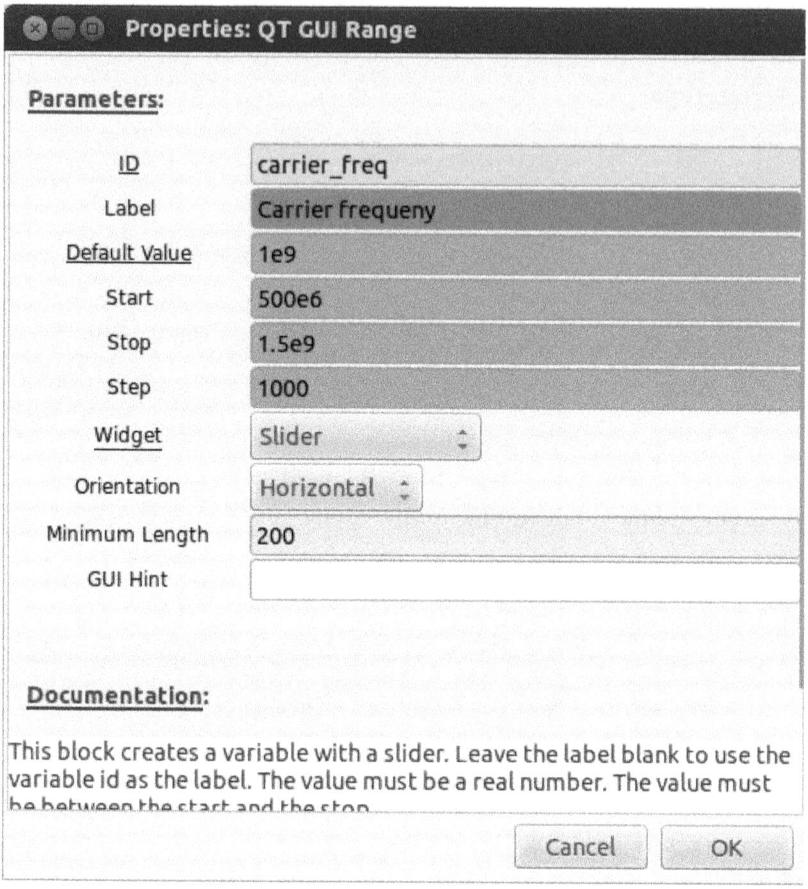

FIGURE A2.5.19
Window showing updated properties of QT GUI range block.

Step 16: From the category **Amitec**, add **Amitec Source**, change
Device Arguments as **amitec** and change Parameters as shown
in Figure A2.5.20.

Step 17: Arrange all the blocks logically connect them as per the
flow-graph as shown in Figure A2.5.21.

Step 18: Execute the flow-graph by pressing F6.

Step 19: Go to tab1 and in at the top of the FLL Bandwidth Range
enter FLL Bandwidth value as 62.8 m and press ENTER.

Step 20: Enter Timing Loop BW as 62.8 m and press ENTER.

Step 21: Enter Costas Loop BW as 62.8 m and press ENTER. This will
display a constellation plot as shown in Figure A2.5.22. Also see
tab3 for demodulated signal.

Properties: Amitec Source

Parameters:

ID	osmosdr_source_0
Output Type	Complex float32
Device Arguments	amitec
Num Channels	1
Sample Rate (sps)	samp_rate
Ch0: Frequency (Hz)	Carrier_freq
Ch0: Freq. Corr. (ppm)	0
Ch0: DC Offset Mode	Off
Ch0: IQ Balance Mode	Off
Ch0: Gain Mode	Manual
Ch0: RF Gain (dB)	0
Ch0: IF Gain (dB)	0
Ch0: BB Gain (dB)	0
Ch0: Antenna	
Ch0: Bandwidth (Hz)	0
Core Affinity	

Cancel   OK

FIGURE A2.5.20
Window showing updated properties of Amitec Source.

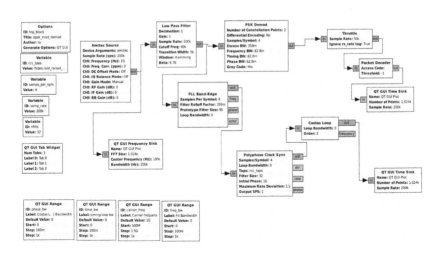

FIGURE A2.5.21
Flow-graph for BPSK demodulation in the receiver side.

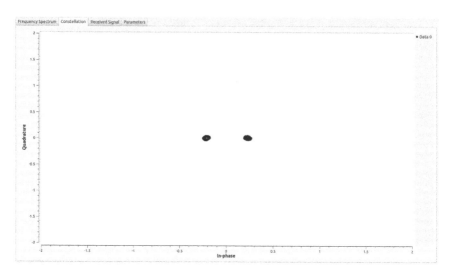

FIGURE A2.5.22
Constellation demodulated BPSK modulation.

5.3 TROUBLESHOOTING TIPS

1. Check that there are no errors in terminal window and ensure that the flow-graph is actually executed. It is possible that ranges will show up even if flow-graph is not running.
2. Go through the Hardware Setup document and follow it correctly.
3. If one observes any error in terminal window saying "FATAL ERROR...," then disconnect SDR04 from machine and connect again. Wait for three LEDs to glow again.
4. Check that the flow-graph was made correctly. Even a mistake in single parameter or block will give ambiguous output which is hard to troubleshoot.
5. Most common errors are made in frequency range and sampling rate. Remember that frequency range should be between 300e6 and 3.8e9 Hz. Minimum sampling rate should be 200,000.

Index